Scientific Thought and Research Methodology

Concepts, Principles, Philosophy of Science, and Ontological Dimension

Aydin Beraha
Cankiri Karatekin University, Turkey

Series in Philosophy of Science
Vernon Press

www.vernonpress.com

In the Americas:
Vernon Press
1000 N West Street, Suite 1200,
Wilmington, Delaware 19801
United States

In the rest of the world:
Vernon Press
C/Sancti Espiritu 17,
Malaga, 29006
Spain

Series in Philosophy of Science

Library of Congress Control Number: 2024944715

ISBN: 979-8-8819-0046-5

Cover design by Vernon Press. Image: Estancia del Sello (Escuela de Atenas).jpg, Raphael, Public domain, via Wikimedia Commons.

TABLE OF CONTENTS

LIST OF FIGURES AND TABLES

Figures

Tables

PREFACE

About This Book

The scientific research process contains many uncertainties for young researchers. Moreover, the terms, concepts, theories, and techniques that researchers encounter can sometimes seem confusing to them. This confusion can be overcome by reaching a state of scientific thought. This book aims to introduce the basic concepts, principles, and research methodology of science to researchers and science enthusiasts pursuing their undergraduate and graduate education.

More specifically, my purpose in writing this book is to present topics that are not commonly covered under the assumption that they are generally known in research methodology books (which contain technical knowledge) that focus more on the use of analysis programs (such as SPSS, Minitab, SAS for analysis, etc.). This book aims to familiarize the reader with **scientific thought** through topics such as the *philosophy of science, logic,* and *methodological or other related concepts,* which constitute the background of and are necessary for the pursuit of social science.

In this book, I describe statistical concepts and related terminology within the context of methodology. In short, this is not a statistics book. Moreover, I do not cover statistical package programs. In this regard, please note that this is an entry-level research methodology book. Namely, the book can serve as a preliminary resource before learning about technical infrastructure (i.e., data collection tools or statistical package programs.).

Let us consider why researchers should pursue scientific thought and learn methodological concepts and the basic assumptions of some approaches of the philosophy of science to scientific research before a statistical package program: Knowledge of the scientific method not only involves ways to collect data and conduct analysis, but also helps one to correctly handle a research study, ask a proper question, define the research problem, and develop a research model. For this reason, before using analysis programs, researchers should develop *a state of scientific mind*; that is, they must understand the systematic nature of science-making which covers *principles, standards, methods,* and *theories.*

In this book, I discuss the basic philosophy of science necessary for researchers without delving excessively into philosophical discussion. I also offer examples from the social and natural sciences (without overwhelming the reader with theoretical details) to make it easier for the reader to understand some concepts related to scientific research. Furthermore, this book, even if it aims to bring readers to a state of scientific thought, has constraints, as mentioned above, concerning philosophical debates and the introduction of scientific concepts and notions. Sometimes, only providing a few sentences about scientific concepts also involves the risk of misunderstanding by readers. To avoid this risk, I strongly encourage readers, when they feel confused regarding any concept, to read further. Additionally, please keep in mind that this book is not a multi-purpose source for scientific methods.

This book is particularly aimed at young social scientists. The purpose of this book is to prepare them for scientific thought while they are in the process of learning scientific methods. Keep in mind that some scientific methods are field-specific! There is no single (multi-purpose) scientific method that fits all fields and subjects.

How to Read This Book

Consistent and appropriate use of concepts is very important in academic literature. Since this book is designed as a general source for the field of social sciences and there may be differences in the topics and methods examined by each field, it is not easy to achieve unity of language. However, I chose to use a group of terms to express the research topics in a comprehensive framework. In this book, I frequently repeat the trio of terms *facts*,[1] *objects*, and *events* to indicate the general subjects of research studies from all social scientific fields.

This book follows particular concepts related to scientific thought. Initially, I introduce the concept of science and scientific fields within a general framework. Furthermore, I present alternative perspectives on science according to its purpose or function in Chapter 1.

Chapter 2 elucidates fundamental subjects requisite for the preliminary phases of scientific inquiry. A short introduction to classical and modern approaches to the philosophy of science is given. For that purpose, I provide general information about the **philosophy of science** and **logic**. It is important to acknowledge that this book adopts a deliberate strategy of simplifying certain concepts and topics to accommodate the breadth of its coverage. However, this approach inevitably entails a degree of surface-level treatment in order to provide an accessible overview of numerous subjects. However, I advise young researchers to study these topics more broadly, especially the philosophy of

[1] *Fact* is a controversial term in philosophy. However, in this book, I use *fact* as an entity to refer to factual realities. *Facts* can entail causal truths such as "Caesar died because Brutus stabbed him," general truths such as death, or particular events such as when Hume dies (Mellor, 2005; Oliver, 2005).

science (and, indeed, philosophy as such). The given content on logic is of a basic level in this book. I remind readers of the basic fundamental principles of logic that most of them probably already acquired in their high school education. In this chapter I guide the reader on how to apply the logical principles in reasoning.

Chapter 2 also presents **epistemology**, which is the science of knowledge. Under the heading of epistemology, standards and verification of valid knowledge are discussed. **Scientific ethics** is also the subject that shares the ethical principles of conducting pursuing science whilst distinguishing field-specific ethical problems. I encourage readers to obtain more detailed information about ethical studies, as each field has its own ethical standards from field-specific sources.

I examine the **ontology** of the scientific research subject in Chapter 3. This strange word means we will be making ourselves aware that we study subjects and variables which exist. I believe that once ontological basics are understood, it becomes easy to handle the variables properly because we comprehend, distinguish, and interpret subjects and variables given the fundamentals of ontology. Ontological dimensions include the *properties* of studied facts, objects, or events, and the *relatedness* of studied subjects in terms of context and operations. In other words, the entity dimension, properties dimension of the subject, and relatedness dimension of the subject in Chapter 3 are the *entity*-related headings.

In Chapter 4, the **human dimension** is examined under two subheadings – those cognitive and technical dimensions. The cognitive dimension and technical dimensions have been considered in the scientific methodology-related section since the human mind (via reasoning) and technical tools (use of language, integrity of hypothesis and research model and providing objectivity of scientific research) should be used correctly in pursuing science. The cognitive concepts presented in Chapter 4 may seem controversial. The main intention of including the cognitive dimension is to illuminate readers about different ways of thinking because there are many ways of thinking in accordance with purpose or need. For that purpose, this chapter presents the cognitive operations of the human mind and introduces how those operations can be achieved effectively. It should be noted that the definitions cognitive-operations-related are compiled in dictionaries and there is no philosophical consensus on those definitions. The purpose of the presentation of cognitive operations is to provide an understanding of their distinctive features.

I have also employed the basic assumptions of *general system theory* as a guide for a better understanding of how to conceptualize scientific research subjects in Chapter 4.

The position of some headings, such as general system theory, may be seen as controversial by some scholars in the philosophy of science. Or readers might ask why I have not explored general system theory in Chapter 3 under the concept of ontology. The reason is that general system theory is a concept created according to the way of conceptualization of the human mind. Further, under the ontology heading, I also explore the concept of randomness, which forms the basis of General System Theory and Chaos Theory. These theories originate from natural sciences and are used in many social science fields because they help understand some social phenomena and events.

The **technical dimension**, in Chapter 4, contains titles such as *objectivity*, *semantics*, and the *use of language*. Semantics and use of language are language-related topics we will cover. I also believe it is essential to choose the right *terms* for concepts in science. This is a quality and technical standard. Sometimes we cannot be aware of the fact that the words we choose to use do not exactly correspond to the concept we want to convey in real life. Or sometimes, two words/groups of words represent the same meaning in our minds, and we cannot see the subtle details of the meaning. This section of the chapter aims to show readers the world of linguistic concepts and other standards related to the pursuit of science but does not enter philosophical discussions of semantics in terms of the philosophy of science.

The use of language is important because you must be able to express your findings well. Thus, rhetoric is crucial. Sometimes, you will be expected to portray the essence of your findings in a few sentences; sometimes, you will need to provide a powerful depiction like a literary writer. In both cases, you should harness rhetoric to use language effectively. Further, findings and conclusions must be put into words accurately and successfully. In doing so, explaining the vital aspects of the subject matter under study may require detailed explanations and evaluations based on different contexts. On the other hand, at a scientific congress and conferences, you will likely present your conclusions, written on dozens of pages, in a mere few minutes. Naturally, the subject matter of the research and the essence of your findings and inferences should be expressed in a limited number of sentences. Scientific writing is like going back and forth between a literary portrayal of the development of an event to the finest details and summarizing the day in one sentence.

In the section on technical dimension we will also look at *structure/integrity* and *objectivity/subjectivity*. Those topics are included on account of their relation to our human features. Research studies necessitate providing the integration of research models with studied theory and being objective in pursuing science.

I present some concepts related to scientific methodology in Chapter 5. The concepts presented in Chapter 5 are vital to comprehending basic notions regarding scientific research. Some concepts may seem close to each other at first glance (such as information and knowledge) and can be used interchangeably. However, there are differences between scientific research concepts, sometimes at a nuanced level, and if they are understood correctly, they can be used in the right context.

I discuss **types of**, and the **process of, scientific research** in Chapters 6 and 7, respectively. In Chapter 6, qualitative, quantitative, and mixed research methods are presented. In Chapter 7, the scientific research process is introduced from determining the research topic to the reporting phase. In Chapter 8, I cover special topics such as *science, ignorance*, the *art of deception*, and *pseudoscience* that some readers might find particularly interesting.

Finally, I have organized the glossary in the last part of the book a little differently from traditional science books; I include some concepts, terms, and even some scientific jargon that are considered important for scientific thought, even if they are not directly related to scientific methodology.

Researchers, especially young scientists who are interested in science and wish to pursue a career in this field, should be equipped with additional qualities. One of them is *versatility*, which is critical because diverse interests improve people's comprehension of facts and events.

Where Are We in the Knowledge Era?

We live in the knowledge era. However, this does not eliminate our need to produce new knowledge. We engage in science not only to satisfy our curiosity also to improve our ways of doing business, to enhance our working life, to make our methods of travel safer, to live more comfortably, to produce better medicines and more effective medical approaches to be healthier, to develop and accelerate our ways of communication, and even to decide how to dress before going out. But maybe more importantly, we come to know ourselves and our environment better through new knowledge. We always need to produce and use knowledge for a purpose, and this will continue. As to what makes scientific knowledge so critical is that sometimes we entrust our lives to it.

Let us not forget to mention the most crucial thing: We must understand the value of knowledge. Francis Bacon once said, "***knowledge is power.***" Such striking words may be effective in showing the importance of knowledge. While engaging with science, our aim is not always to produce extremely striking results (such as a revolutionary new teaching technique in education). Each small piece of knowledge produced may contribute to the accumulation of wisdom and progress in each field. Hence, every research finding need not make big statements. I only want to draw attention to understanding the significance of knowledge here. Of course, each piece of knowledge created by us is not for improvement or for a particular purpose. But social progress depends on generating knowledge. Bacon, in his famous aphorism, emphasized the improvement of human life. Using the right knowledge in the right way saves lives. A lack of required knowledge or knowledge used incorrectly makes people's lives miserable. It is that simple.

Returning to the previous paragraph, we must not see knowledge only as a source of power. That is right, in a sense. On the other hand, we must consider knowledge as a source of well-being for the life and happiness of humankind. Aphorisms such as "***knowledge is power***" sound more political nowadays, and I see, personally, knowledge as a source of progress in human well-being, not only as a source of political power.

We have shaped our modern life in every way via humanity's accumulation of knowledge. We are so accustomed to this situation that the knowledge behind every blessing from which we benefit is no longer visible to our eyes. However, in ancient times, knowledge was scarce and written sources were rare. We can better understand this when we realize that many lives have faded due to a lack of small pieces of knowledge (i.e., poisonous or medicinal plants) throughout the history of humanity; many people have survived due to having those pieces of knowledge and living in an ocean of information, which makes our lives easier today, with such great wealth and blessings for humanity.

Our ability to produce knowledge at a dizzying pace looks gorgeous. Its natural conclusion is the rapid aging of some information. From this angle, we can also say that we live in an information dump containing incorrect and nonfunctional information. This is because each piece of knowledge we obtain is not necessarily correct, functional (for human benefit), or up-to-date. The correct use of knowledge is also vital. Therefore, we also need both general and scientific ethics to guide us in how to use our knowledge.

In the knowledge era, we are, at the same time, living in an ocean of incorrect information. All information surrounding us is not correct. False information can be produced for manipulation, which is also discussed in this book. Besides, information pollution is one of the up-to-date problems of modern life. There are many websites that claim to offer scientific knowledge. Sometimes well-known news sources share deceptive information or include some pseudo-scientific findings without seeking expert opinion. Moreover, most of the knowledge produced today is temporary in nature (yes, information is getting old, but recording and archiving can be functional), and it can only be functional under particular conditions. Only a small part of the information we produce qualifies as permanent; this is generally the case in the natural sciences. Of course, natural sciences also produce temporary knowledge. For example, meteorology makes estimations of daily weather. Hence, the knowledge meteorology produces is temporary. Since much of the knowledge generated in the social sciences reflects the reality of certain periods and conditions, it is mostly temporary. Thus, knowledge produced in the social sciences has a bad habit of getting old fast. For example, consumer expectations are constantly changing under the influence of social media and fashion trends, labor market characteristics vary from generation to generation, and new social realities require new institutions. The education system is constantly changing everywhere. Today, the fact that we have a fabricated education system (the content may differ, but the systematic context is the same almost everywhere) left over from the industrial revolution is being questioned, and some educational scientists seek to change it.

In the present day, in addition to producing knowledge, it is critical to compile existing information (such as through time series analysis) or to reprocess it with different information (through synthesizing) to make it more useful. In short, we encounter processes such as synthesizing and re-processing information with different components and in different contexts. Further, concepts and applications such as big data[2], data mining, and Web browser analytics are entering our lives. The better organization and visualization of knowledge used in education is increasingly gaining importance, improving the quality of education in the process. As the use of infographics has become widespread, it has become easier to present and to gain knowledge.

Scientific knowledge is only one of the many types of knowledge that shape and guide our daily and professional lives—such as *everyday knowledge, philosophical knowledge, art knowledge,* and *technical knowledge*—but perhaps the most important one. All types of knowledge mentioned above appear to some extent as extensions of *scientific knowledge.*

Scientific knowledge production requires recognition of scientific methods. When we say *method,* the first thing that comes to mind is to use predefined (and also consistent) ways and techniques to be followed in order to answer a scientific question or solve a problem. Scientific methods provide consistency and an adherence to standards, which are important because the purpose of scientific methods is to produce accurate knowledge.

---

[2] *Big data* refers to large digital datasets that arose from advances in technological capacity and analytical methods. These datasets cover "everything within a particular field (e.g. utility records) or platform" (e.g., Twitter or Facebook) so that "states and corporations are able to collect, store, and process more data than ever before". The digital nature of these data creates "potentials for data mining and data linking, allowing connections to be made between diverse data" (D'ignazio, 2019: 2; Tinati et al., 2014: 664).

Chapter 1

UNDERSTANDING THE SCIENCE

Before starting to read the chapter

Science is one of the long-term pursuits of human history. We do not yet know where we are and where science will go in the future. We are still debating our scientific methods. New branches of science emerge over the decades. Despite all this turmoil, we continue to engage in one of the most meaningful human pursuits.

This chapter introduces a brief overview of the *concept of science*. It explains what science is, what kinds of activities are considered science, and what kinds of features in a researcher are necessary for conducting good science. Then, the chapter introduces general overarching branches of science. Lastly, the chapter presents a classification of science according to the nature of pursuing science.

SCIENCE

What is Science?

Science is one of the most crucial pursuits of human progress. This concept—which is not easy to attribute a beginning to—brings to mind the first inventions when viewed through a historical lens. Perhaps the invention of the wheel can be seen as a milestone, yet it seems as if the wheel is more of a technological development. Yes, we will also mention the relationship between science and technology. Would it be wrong to think that science started with philosophy? Science has a philosophical dimension, and we cannot deny the contribution of Ancient Greek philosophy, especially to science as a science. What we call today science in the 19th century came out of natural philosophy, which is discussed in detail in Chapter 2.

Returning to the word *science*, when we decide to do brief research on books and on the internet, we come across a wide variety of definitions. For example, science is defined by the Merriam-Webster Dictionary (n.d.) as "Science is knowledge or a system of knowledge covering general truths or the operation of general laws especially as obtained and tested through scientific method." The Science Council (n.d.) defines it as "Science is the pursuit and application of knowledge and understanding of the natural and social world following a systematic methodology based on evidence." Based on those definitions, we can say that science is the job of producing knowledge about facts. To explain

> **"Science is organized knowledge."**
> (Herbert Spencer 2008)

it in terms of more specific features, science can also be defined as "the sum of activities to understand world, events, and phenomena by using systematical methods." If we pay attention to the emphasis in the definitions, **knowledge** is the focus of science, and the **activities** to produce knowledge are highlighted.

From the standpoint of epistemology, which is a special field that deals with knowledge, and scientific methods, all kinds of facts, objects, and events that exist can be the subjects of science. This means that our universe, the Earth, and the social environment in which we live contain numerous facts, objects, and events perceived and measured by us. Science chases all that can be observed and measured.

If we look at science in terms of activities, the systematic search for knowledge appears. The systematic search activities for knowledge start with questions on a logical basis, continue with creating a research design based on the theoretical framework and scientific methods, and conclude by making inferences again on a logical basis. Briefly, systematically seeking and producing knowledge related to **logic**, the **philosophy of science**, and scientific **methods,** which are discussed under related titles in detail.

Science deals with (or reveals) *components* (Brigandt and Love, 2023), *features* (Kim, 2005), *processes* (Fraenkel, Wallen, and Hyun, 2011), *effects, relatedness* (Fraenkel, Wallen, and Hyun, 2011), *randomness* (Dembski, 2005), or the *temporality* of the subject(s) discussed. Thus, it is possible to look at each subject examined in different contexts. For example, in the analysis of historical events, there are many contexts, such as chronological (i.e., the timeline of a set of events); political and social causes (social factors); stages of events; and political, economic, and social consequences (exposed effects) (Tanaka, 2019). Scientific research can focus on one or more of these contexts.

Science includes the following activities:

- ⇨ *Defines* and/or *explains* concepts, structures, components, features, classes, actions, behaviors, processes, functions, limitations, effects, causality, etc.
- ⇨ *Classifies* species, structures, and models etc.
- ⇨ *Analyzes* the data
- ⇨ *Makes inferences* for the present status of fact or population and/or makes *predictions* for the future of fact or population

All these activities of science can be *explanatory, confirmatory* or *predictive.*

What has no place in science? *Miracles* and *mysteries* have no place in science. **Miracles** violate the laws of nature. **Mysteries** are not-understood phenomena, and are nothing but a lack of knowledge. For example, an unidentified aerial phenomenon (UAP) is not a mystery but a lack of knowledge about a phenomenon. On the other hand, a miracle-look event may be an experience that can be reduced to probability. An event (which is often considered a miracle) occurs due to a series of low probability, pioneering conditions, and events that can be explained by reducing them to some calculation of probability. A positive outcome is a prerequisite for a miracle. In unfortunate events, the concept of *bad fate* replaces the miracle. For example, the death of everyone except one passenger in an airplane accident forces most people to consider that a miracle may have ensured the passenger's survival. However, a number of physical factors may have caused the passenger to survive by chance.

Every not-understood phenomenon and event is a mystery to humans. This is often the result of people linking phenomena or events they cannot explain to uncertainties. For example, in the past, people did not know how natural events occurred or the natural forces behind them. Therefore, they associated each natural event with a mysterious power (often described as a particular god who was the owner of the mysterious power). So-called "mysterious" forces of nature were attributed to sea gods when tsunamis happened, to sky gods (or gods of lightning) during floods, and to underground gods when earthquakes struck. Even today, there are many facts and events whose dynamics we do not understand fully. The Big Bang is just one of them.

In science, **intuitive prediction** and **faith** have no place because science is concerned with evidence and progress via proof and verification. Intuition can only be a guide for scientists in some cases. Please note that we are not dealing here with estimations! Sometimes scientists act based on their intuition if there is not much information to guide them about where to look for more details on a particular subject.

Proof or verification of some assertions cannot happen by itself. As in the field of law, there is no self-testimony in science. Personal skepticism is not verification. For example, the following proposition has no place in science: "*This is unlikely to be so, so this is not true.*" There is no evidence put forward in this example; there is only personal skepticism.

What Does a Scientist or Researcher Do?

Whether one is a scientist or a researcher, every individual can produce knowledge. Initially, let us explain why we separate the concepts of researcher and scientist. While a scientist is a researcher by profession, a researcher is not necessarily a scientist. Because there can be researchers in fields other than science. For example, the researchers of a company doing market research are not scientists, but the methods and tools they use are scientific. This example primarily points out that there are differences in motivation and goals because they progress on different career paths. On the other hand, the methods and tools used overlap. Even if their methods and tools used overlap, yet, researchers as scientists have distinctive roles within the science community. As scientists, researchers need to integrate their scientific research with the theoretical knowledge of the field in which they work. This points to the contextual aspect of research for scientific purposes and distinguishes it from, for example, research for commercial purposes.

The activities that a scientist or researcher performs while generating knowledge include collecting new information (or gathering available information) and analyzing, or synthesizing information. The production processes of knowledge conducted by scientists or researchers are expected to be systematic.

Briefly, a researcher:

⇨ Asks questions,
⇨ Searches for information (by collecting data), and
⇨ Performs analyses and inferences.

Dealing with science requires certain qualities. Above all, **objectivity** is the basic principle of science. People can be biased due to their personal, cultural, and historical contexts. Even so, the researcher should separate her/his worldview from science as much as possible[1]. On the other hand, there may be differences in the level of objectivity depending on the field of science. In addition, the measurement instrument must be objective. Objectivity standard of measurement instrument is related to *validity* and *reliability* concepts which are discussed in Chapter 7 elaborately. There is objectivity if only two separate research studies, conducted on the same unit of analysis of the same subject, reach the same results (provided that there is no methodological error) when evaluating the findings of a measurement (even if the studies used different measurement instruments) (Creswell and Creswell, 2017; Maxwell 2013; Outhwaite and Turner, 2007).

Descartes was dedicated to reaching the truth and believed that a discovered truth would also produce excellent consequences for practice (Grene, 1999). Descartes placed great importance on being **skeptical** in science in terms of method and employed doubt as the basis of epistemology (Sankey, 2012). Descartes proposed the principle that any given perceptual experience might be false given that example "*there are never any sure signs by means of which being awake can be distinguished from being asleep.*" This argument originated the idea that "if for any perceptual experience, it is possible that there should

[1] According to some scholars, it is challenging to achieve utter objectivity and to remain free of bias, especially in the social sciences.

be a non-veridical but phenomenally indistinguishable experience, and then the first perceptual experience cannot be trusted to be veridical" (Bermudez, 2008: 14). However, Descartes's skepticism was radical and his concerns about skepticism were not practical due to he opened a questioning path of methodology for *the existence of the external world.* Since this radical skepticism led him to be skeptical against skepticism also, his method restricted his own skepticism due to everyday practice was not compatible with such hyperbolic skepticism (Grene, 1999). This argument is based on Cartesian Skepticism, which indicates one's being skeptical about her/his sense perceptions (Grene, 1999). This is the main difference between Descartes and the Ancient Greek philosophers. Instead of asking such questions regarding *the existence of the external world,* Ancient Greek philosophers preferred to stay within the practice of daily life in their philosophical discussions (Grene, 1999).

A mind that thinks skeptically is very important for scientific progress. The essence of science necessitates being skeptical when there is not adequate evidence (Rothchild, 2006: 1) or verification. An understanding based on a dogmatic, unquestioning view of truth will fail to generate scientific knowledge.

Another significant property of engaging with science is **open-mindedness**. Scientific knowledge can be produced through the efforts of an open mind accompanied by skepticism. This standard is closely related to those of objectivity and skepticism mentioned above for the same reason that "observable facts count rather than one's personal views" (Mitchell and Jolley, 2009).

The starting point of science is **curiosity,** in many cases during human history. Even today, engaging with science sometimes starts with curiosity. This may seem like a nostalgic feature when pursuing science today, when academic or professional career plans are privileged, and the main reason for being involved in science is to do research. In other words, if people are primarily motivated by their careers, we would probably be right about curiosity's being a nostalgic feature. However, even if people pursue science for career purposes, this does not prevent researchers from acting with a sense of curiosity according to their expertise or interests. Indeed, it is worthwhile to keep the curiosity of young researchers and science lovers alive.

Before producing new knowledge, we should seek to answer the questions in our minds from previous studies. Hence, it is necessary to read prior studies to glean information about the questions to which we seek answers. We must not forget that, before starting research, we need theoretical knowledge about the topic at hand, which is revealed by delving into past findings. Theoretical knowledge also provides assumptions about the research subject, and assigns the standards for the initial conditions of the research. This leads us to original research activities for knowledge production. It is then possible to **search for**

new knowledge. This is original knowledge that we will provide through our own investigation.

The distinguishing feature of science is that it employs careful, systematic methods that enable its claims to be based on **evidence**. Before a researcher begins to do research, she/he must know where to find information and how to gather it. This task requires methodological knowledge. **Methodological knowledge** encompasses the activities of searching, collecting, classifying, and analyzing information, and requires one to follow a set of rules and standards.

The correct application of scientific rules and standards, as well as research methodology, ensures that a data collection tool is **reliable** and **valid**. Thus, it is possible for the findings obtained from research to reflect the truth about a particular situation. Rules and standards providing reliability and validity also allow independent studies to confirm each other's findings. In case the subsequent research studies falsify the previous findings this indicates two possible reasons. One possible reason is that the subsequent research study failed to reach the desired reliability and validity standards of the research. The other possible reason is that the results differed for some reasons, such as data processing errors, inferential errors etc. The researcher should determine why the results differed.

Scientists propose several theories and evaluate them in light of evidence gathered through observations and/or experiments. In other words, information obtained from empirical studies in a particular field may lead to the development of theories (induction) in that field. But in most cases, scientific research studies follow available theories in their disciplines and researchers establish their research design in accordance with this theory. Those studies lead to the support of an existing theory (deduction).

Scientific theories comprise the entirety of information and explanations that are put forward in the process of acquiring knowledge. As the amount of empirical information supporting a theory increases, the theory becomes a **scientific law**. Briefly, a theory becomes a **scientific law** as the amount of empirical knowledge that supports a theory increases.

Measurements are useful for building and developing theories. The more qualified measurements a scientific theory is supported by, the stronger it becomes. A qualified measurement (see Chapter 5) contains the following features (Fraenkel, Wallen, and Hyun, 2011; Turner and Outhwaite, 2007):

- ⇨ Research design compatible with the theoretical framework,
- ⇨ Properly selected and sufficiently sized sample selection,
- ⇨ Proper data collection tools [instruments],
- ⇨ Determined limitations

If there are different competing theories in the same field, and if they progress with the accumulation of knowledge, a synthesis can be produced on subjects studied that do not contradict each other, or one theory is counted as valid.

The knowledge produced by past studies corresponds to systematic accumulation and is usually aggregated around several theories in the related area. However, scientific knowledge belonging to a field and gathered around certain theories may not always be sufficient, as Thomas Kuhn (1962) pointed out, a change may occur in the form of a paradigm shift. But in most cases, gaps exist in the knowledge network (called theory) may need to be filled. These gaps are opportunities for new research studies. Questions arise from following these gaps or contradicting findings of previous research. At this stage, it is necessary to ask question(s) about the subject (or professional interest). Scientific activity starts right here by **asking questions**.

Asking questions in science is not a job that can be done just as they come to mind. First, it should have a **logical basis**. Questions that meet both the needs of the research and the requirements of science and logic are valid, which is why not only methodological, but also **theoretical knowledge**, is required. Consequently, given the logic dimension and the need for theoretical knowledge, asking a scientific question requires knowledge of the background because it is not just about asking questions. The point includes:

- ⇨ Asking the right questions (this requires theoretical knowledge and learning from past findings) and
- ⇨ Asking questions correctly (this requires combining methodological and theoretical knowledge with a valid logical basis).

Functional (e.g., solving a social, natural, or technical problem) and rational questions require reasoning (e.g., analytical, inferential, or contextual), as well as a maximum amount of theoretical knowledge on the subject. Asking the right questions requires correct logic and rationality, as well as conceptual and theoretical knowledge on the subject. Do not be afraid of this! Engaging with science should not sound scary. Let us perform an exercise:

Let us imagine that a comet is approaching the Earth and is expected to pass very close to the Earth. After a while, it passes by the Earth and we survive! What might be the first question that comes to mind?

"When is the next transit of the new discovered comet?"

If you are not very familiar with astronomy, this will probably be the most scientific question you can think of. If so, you can rejoice. The author of these lines is as unfamiliar with the subject as you are. Nevertheless, even reaching this question is crucial because it is necessary to consider many concepts related to the subject in order to ask the *right questions* (also the best questions) and in the right way. If we consider this matter, it is very natural for us to want to know the risk that a comet poses to the Earth. By calculating, we can know how much time we have until the next encounter and, if necessary (if the risk is high; that is, the possibility of impact is significant), we can prepare for its return. Therefore, the first, reasonable question that may arise from this risk will be about determining the time when our newly discovered comet may arrive in our neighborhood.[2] Astronomers will probably ask this question as follows:

"What is the orbital period of the new discovered comet?"

If you pay attention, the two questions above are very close to each other. However, the latter question seems more appropriate to astronomy. By the phrase -asking the question in the right way- I mean giving the question a more scientific touch.

When evaluating a topic, the human mind works similarly up to a point, but when field knowledge and expertise come into play, logic execution gains depth and makes it possible to continue to ask more vital questions about the topic. This is exactly what I meant when I said that theoretical knowledge is required to ask the right questions.

We can also ask the following questions regarding our comet problem:[3]

"What is the size/diameter of the comet?"
"What is the impact probability of the comet?"
"What is the impact energy of the comet?"

As you can imagine, the orbit and velocity of the comet are critical to accurately calculating the next visit to our neighborhood. The information on the size (i.e., diameter) seems important (according to my own knowledge of astronomy, which is very limited) in terms of calculating the damage it could cause in the event of a possible impact.

These questions can be diversified with far more scientific questions, but since the expertise exceeds my own knowledge, let us set this subject aside now and continue our exercise with an example from the social sciences.

---

[2] Periodic transitions of asteroids and comets can vary from a few years (Icarus: 1.12 years; https://nssdc.gsfc.nasa.gov/planetary/factsheet/asteroidfact.html) to decades (Halley comet (officially 1P/Haley): 75.9 year; https://ssd.jpl.nasa.gov/tools/sbdb_lookup.html#/?des=1P. This information is important for us to prepare (if the calculated risk is high).

[3] The questions were generated from the study of Shustov and Rykhlova (2009) and content from NASA's Center for Near-Earth Objects Studies (CNEOS). In this context, astronomers will begin to ask more advanced/scientific questions.

"Why are all Asian countries underdeveloped?"

The question may not look wrong, although it creates a strange feeling. If you consider this question rationally problem-free, you can start looking for answers. Your path is long. However, the question gives the impression that it contains a *bias* that is far from scientific. The adjective -all- gives the clue that there is a bias here. Therefore, it is a very wrong question. Prejudice (bias[4]) is an anchor that is closely related to subjectivity (as a problem of objectivity -See Chapter 4), and might hamper objectivity at the stage of asking questions or developing a hypothesis. The question has a starting point that views "*all*" Asian countries as underdeveloped in advance. Further, "*all*" and similar expressions (e.g., everything) are not suitable for use in the social sciences, as they represent a generalization over the entire population. Nevertheless, even if we consider this "*provocative*" question to be valid, this question should clearly be a preliminary one with a logical inference. Thus, let us reformulate our question as follows:

"Are all Asian countries underdeveloped?"

Our question now seems more accurate! Yes, it is undoubtedly more accurate, but we can still see through reasoning, that it is still too advanced to serve as a starting point for the question. We have a critical component (term) in both questions—the term "**underdevelopment**." In order to conduct scientific research, this concept and its indicators (sub-dimensions) must be defined because it is the *keyword*. What do we mean by "underdevelopment"? What are the indicators of "underdevelopment"? In short, before seeking to answer the second question above (which seems more rational than the first), the concept of "underdevelopment," which lies at the core of our research problem, should be understood very well! The reason for our previous statement, "*asking the right questions requires conceptual and theoretical knowledge,*" is probably clearer now.

Therefore, experienced scientists read extensively (as much as possible) about theoretical underpinnings before starting research. Even the nuances between concepts (terms/themes) can change the content, scope, direction, and quality of the research. For example, scholars studying organizational behavior frequently pursue inquiries on motivation subjects. Bullying in the workplace should be studied as a negative factor affecting employee performance because the anticipated effect of bullying in the workplace would be negative on performance. In other words, bullying is not anticipated to cause an increase in the performance of employees. The negative impact of workplace bullying affects both the choice of scale to be used to measure performance and the inferences drawn from the analysis. Since motivational

---

[4] This is also a disadvantage of writing a hypothesis (Fraenkel, Wallen, and Hyun, 2011).

subjects such as workplace bullying are measured at the individual level in the field of organizational behavior, performance also requires measuring at the individual level. Briefly, variable relationships should be established correctly on the basis of logic, and a research design should be made in accordance with the assumed direction of the relationship.

The accepter of our social-science question above seems to be economics. This is partially right because the term "underdevelopment" is associated with both economic and human indicators. Further, "underdevelopment" refers to socio-economic indicators of being poor. If we are not familiar with the term as a concept, we may have a conceptual confusion. For example, we may be confused about the difference between the terms *economic growth* and *human development*. Consequently, "underdevelopment" is both a socio-economic concept and an interdisciplinary subject.[5]

Since our question seems to cover the concept of human development, which is a socio-economic concept, the indicators we will examine are sub-dimensions of human development described by UNDP (United Nations Development Programme), as follows:

- ✓ Life expectancy
- ✓ Education
- ✓ GNI (Gross National Income) per capita

Thus, we have to study the development indicators of each Asian country. The indicators we are going to look at lead us to a guiding question. Now, our guiding question seems to be: "*Are some Asian countries less developed than others?*"

At this stage, we make a *comparison* between countries in which we identify the indicative data. The answer to the above question is simply "*yes*" or "*no.*" After comparing the indicators of development of Asian countries, our above

---

[5] The concept of **economic growth** refers to a country's production and expenditure power. In this regard, it is more related to a country's industrial potential. The concept of **human development** refers to progress in terms of a country's human and economic indicators. In this respect, it is a more comprehensive concept (including economic growth). In other words, economic growth encompasses a country's production power (based on various industries) expressed in terms of GDP (Gross Domestic Product); the concept of human development is related to indicators listed under the human development index (HDI-which is used by the United Nations Development Programme), and contains indicators such as a country's literacy rate, schooling rate, quality of education, infant mortality rate, number of people per doctor, quality of health services, purchasing power parity, and women's employment. Even if human development is a broader (socioeconomic, and thus interdisciplinary) concept, there is as well a two-way relationship between economic growth and human development (Ranis, Stewart, and Ramirez, 2000; Ranis, 2004).

question (with the answer being "yes") is followed by: "*Which Asian countries are less developed than others?*"[6]

Suppose we found the answers. They will show us that our first question was "wrong." The answer to our second question will simply be "no." Hence, not all Asian countries are underdeveloped. Indeed, asking a wrong question seems to fail more than a mere question whose answer is "no." since a wrong question directs you to a dead-end but a mere question directs you to a simple conclusion.

An important point should be noted here. We can only determine underdevelopment by comparing development standards. Hence, we need to see the contrast so that we can tell who has been left behind according to which indicators. Eventually, we should have values of development standards of Asian countries that would be relative rather than absolute as variables are in many social science fields. For this reason, while making a comparison, a relative (versus absolute) result is produced in social sciences.

In addition, the questions may raise several derivative questions, which can also be dimensional. In other words, new questions are dimensionally positioned when they follow and explain the first question.

1. How far/close are Country A's human **development indicators** from the top 5 countries with the best performance in Asia?[7]
    - 1.1. How far are Country A's **economic indicators** from the average of the top 5 countries?
    - 1.2. How far are Country A's **demographic indicators** from the top 5 countries with the best performance?
    - 1.3. How far are Country A's **education indicators** from the top 5 countries with the best performance?
    - 1.4. How far are Country A's **public health indicators** from the top 5 countries with the best performance?
    - 1.5. How far are Country A's **transport infrastructure indicators** from the top 5 countries with the best performance?

---

[6] It should also be noted here that the subject discussed here is too broad for a socio-economic research. Subjects of study in the social sciences are commonly more specific.

[7] Regarding this question, we are supposed to determine five of the most developed countries in Asia. Afterward, we should calculate the average of five countries for each development indicator. At the final stage, we can make a comparison to find answers to sub-questions.

While doing all this, a researcher should decide *what information* she/he wants to find out and *how* to do so.

- ⇨ A **"what"** question indicates which factor(s) produced results. A "what" question has a *descriptive* feature. For example, "What happened?" (Gay, Mills, and Airasian, 2011).
- ⇨ The question of "**how**" refers to how a cause (factor) reveals a result; in other words, the way the effect occurs. This "how" question has a *descriptive* feature as well (Rescher, 2003).
- ⇨ A "**why**" question refers to the characteristics of the cause (factor) creating an outcome. In other words, it concerns several characteristics of the factor(s) that bring about a result. The characteristics (also referring to conditions) or deeper factors (hence the chain of factors) of the revealed effect/cause may include the answer to why a particular outcome happens. Hence, a "why" question has an *explanatory* feature (Gay, Mills, and Airasian, 2011).

A scientific *description* is related to *what* and *how* questions about nature and social issues. A scientific *explanation* is related to a *why* question about the effect or result (Rescher, 2003). *Why* questions generally (but not always) call for an experimental research design, while *what* questions generally (but not always) require a non-experimental research design (Mitchell and Jolley, 2009).

Examples for the *why* questions (Chigbu, 2019):

- ⇨ "**Why** are women poorer than men?" or
- ⇨ "**Why** do developing countries suffer more economic crises than developed ones?"

Example for the *what* questions (Chigbu, 2019):

- ⇨ "**What** are the sources (or factors or determinants) of high inflation[8] in Country A?"

Let us suppose we are seeking to answer the question above (regarding high inflation), and let us assume the answer is "devaluation." Thus, "What caused the increase in inflation?" The answer to this question can lead the researcher to the devaluation (factor/independent variable). Alternatively, the source of inflation may be the current level of demand (factor/independent variable).

Put another way, devaluation (the first independent variable) and demand (the second independent variable) may have led to an increase in inflation

---

[8] High inflation can have many causes such as monetary expansion, the cutoff of foreign loans, investments due to exchange rate shocks, an increase in government spending, supply shocks, and output gaps (demand-driven inflation e.g., the oil crisis of 1973) (Ball, 1993; Dornbusch and Reynoso, 1989; Smith, 2005).

together. In the present case, the relative distribution of the resulting effect between two factors (independent variables) would also make sense. Thus, the effect weights of both independent variables must be determined.

"**How** did this effect occur?" When seeking to answer the question, we need to determine how the effect occurs. In the first example, the probability of the effect is that the foreign currency deficit caused devaluation, while devaluation triggered an increase in inflation. Now we have a new question: "Why did the foreign exchange deficit occur?" The answer to this question is as follows: The foreign exchange deficit was caused by the fact that the foreign currency arriving in the country through exports and services has a lower value than the foreign currency derived from imports and services. Here, the factor's dependence on the other factors (the chain of factors) is explained. Let us take a look at another example below.

"*What factors cause cultures to develop*?" Let us suppose that the answer to the question is economic activities. Our conclusion would be: Different kinds of economic activities (the answer to the question of **what**) caused cultures to develop.

"How do economic activities affect the development of culture?". Needs enable the development of new behavioral practices (the answer to the question of **how**), and these behavioral practices alter into cultural features that characterize society over time. Concretely, the needs of societies close to the poles (protection against the cold and food supply, etc.) result in appropriate economic practices (i.e., protective clothing production, hunting in snow and ice) with these climate conditions. In this respect, it points to the question of how a culture evolves.

In some cases, question words may need to follow one another. For example, researchers may need to answer *what* questions to bring factors (causal conditions) into the open, and then *how* questions to uncover the ways in which effects occur. In the context of philosophy and logic, there are basic questions –regarding *what* and *how* questions- specific to various fields that reflect the nature of those fields. Some of these are (Smith, 2018):

- ⇨ The question of ontology is "What is it?"
- ⇨ The question of phenomenology is "How do we experience it?"
- ⇨ The question of logic is "How can we reason?"
- ⇨ The question of epistemology is "How do we know?"
- ⇨ The question of ethics is "How should we act?"

These examples are probably sufficient for us to understand the nuances required to ask questions. Let us set the subject aside for now.

The distinctive feature of science is that it uses careful, systematic methods that base its claims on **evidence**. Therefore, researchers need **methodological knowledge**, which includes seeking, collecting, sorting, and analyzing information, and requires following a set of rules and standards. The correct use of research methodology, as well as rules and standards, ensures that a data collection tool is **valid** and **reliable**. Thus, it is possible for the findings obtained from research to reflect a real situation. A researcher also must know where to find information and how to collect it before initiating a survey. Rules and standards also allow research to be conducted independently so that findings from different studies can be verified by following the same rules, or falsified by revealing the reasons.

Scientists propose several theories and evaluate them in light of evidence collected through observations and/or experiments. In other words, the information drawn from an empirical investigation in a certain area leads to the development (through induction) of theories in that area. In this context, a **scientific theory** is the collection of pieces of knowledge that are determined in the process of obtaining information whose validity and reliability are determined by scientific standards and have internal consistency.

Measurements (or testing) are useful for creating and developing theories. A scientific theory is stronger if it is supported by a larger number of qualified measurements. This also indicates that scientific research should follow procedures to collect and analyze data in order to be replicable and verifiable (Chigbu, 2019).

> **"I am among those who think that science has great beauty. ... A scientist in his laboratory is not only a technician but also a child placed in front of natural phenomena which impresses him like a fairy tale.."**
> (Marie Curie 1937)

If there are different theories competing in the same field and they are progressing with their own knowledge, a synthesis can be produced on the subjects where they do not contradict. If empirical studies generate results that confirm opposite hypotheses about the same subject, an empirical study should seek to identify a deficiency (a factor or condition may be lacking) and/or errors (the data collection or analysis method may be inaccurate).

Science also has practical goals such as *prediction* and *control.* Science provides us with successful estimations about nature and social issues in terms of *prediction.* Science also allows us to control nature in a desired direction (Rescher, 2003). For example, we make estimations about weather conditions for next week, or market growth for the next years. We take water under control via channels and dams and manage it. Our second example brings us to a particular type of science called *applied science* which refers to the implementation of a scientific idea and knowledge (see the title Basic Sciences and Applied Sciences). If we summarize what we have examined so far, we can divide scientific activities into three categories: *theory development, research,* and *application.*

BRANCHES OF SCIENCE

This section may not seem very important at first glance. The main purpose of writing this section is to remind the reader that there is dynamism in disciplines (or fields), and to provide information in light of recent developments. Science is prevalently divided into two main fields: the **natural sciences** and the **social sciences**. Many undergraduate and graduate programs also give a place to another main field called **humanities,** which is the study of humans and includes fields such as language, literature, art, etc.

The natural sciences are further divided into two areas: the *physical science* and *life science.* The physical science contains physics, chemistry, astronomy, meteorology. Life science contains biology, biochemistry, botany etc. Engineering is related to technology and is classified under the applied sciences. The applied sciences also contain medicine and health care. Fields such as mathematics and logic are grouped under formal science.

Many authorities make different classifications. If we consider the *Encyclopedia Britannica* or a detailed alternative, we can examine the science and technology nomenclature of the United Nations Educational, Scientific and Cultural Organization (UNESCO). The journal *Nature* also presents scientific classifications. Key categorical fields covered by the journal include the *physical sciences, earth and environmental sciences, biological sciences, health sciences,* and *scientific community and society.* We can look at the work of Glänzel and Schubert (2003)[9] for an alternative classification of scientific disciplines.

An important distinction here regards the division of branches (used in the same sense as fields or disciplines in this book) into sub-disciplines (sub-branches) e.g., physics has divisions such as quantum physics, nuclear physics, and astrophysics. As you might expect, the numbers of both disciplines and sub-disciplines were limited in the past compared to the present. In fact, many branches of science that we are familiar with today emerged in the late 19th century and in the first half of the 20th century. Some scientific subjects have become separate disciplines (branches) by moving away from the field(s) they were originally part of.

---

[9] Glänzel, W., & Schubert, A. (2003). A new classification scheme of scientific fields and subfields designed for scientometric evaluation purposes. Scientometrics, 56(3), 357-367.

Every scientific discipline has sub-disciplines. In addition, scientific disciplines may have mutual fields of study, which are called **interdisciplinary**. In particular, the complex nature of social events underscores the need for interdisciplinary studies; as a result, interest in interdisciplinary studies increased in the 20th century. For example, when we examine the factors behind social events at the individual level, we encounter **social psychology**, or when we aim to explore the nature of economic events in the context of social events, we encounter **socioeconomic** studies.

Some terms, concepts, and modeling of the social sciences and the humanities come directly from the natural sciences (Butz, 1995). This is because the functioning of nature and the functioning of social and human events are based on the same natural laws. We will discuss the relationship between the laws of nature and the functioning of social life in the course of the book.

For example, the **General System Theory** mentioned earlier was originally an approach in the natural sciences. However, the theory has been used to describe social structures, such as industrial systems, or events and facts, such as political movements and trends (Chen and Stroup, 1993). Another example is **Chaos Theory**, which is philosophically old but scientifically relatively new (Butz, 1995). The theory emerged in the natural sciences, but soon began to be used in the social sciences and humanities, such as psychology (Butz, 1995; Kiel and Elliott, 1997). **Population ecology** was once used in the social sciences, similar to General System Theory, to describe social structures such as industrial systems (Hannan and Freeman, 1977). We will discuss these concepts in Chapter 4.

CLASSIFICATION OF SCIENCE

There are various classifications of science according to different purposes, perspectives, and functions. Some of these classes/categories of science are introduced here.

Basic Sciences and Applied Sciences

Some authors use technology and applied science in the same sense. Although there is a subtle difference between them, this similarity is acceptable in terms of technology. Nevertheless, applied science refers to the application of a scientific idea and knowledge, which does not always imply technology. In other words, technology is one type of applied science, but it does not constitute all of applied science. For example, new educational or industrial activities are examples of applied science in which technology is not involved (Bunge, 1966). To embody this distinction, we can turn to the following example: Trying new managerial practices that will increase employees' motivation in light of new knowledge in the field of organizational behavior fits the definition of applied science, but technology may be irrelevant. New motivational tools include applications such as wages, social rights, and participation in decision-making processes, which are not related to technology. Similarly, new health care or educational treatments in the context of new knowledge can be accepted as applied science, but may not include a technological aspect. Hence, applied science refers to the application of scientific knowledge, and may include technology or not. Therefore, it would be more appropriate to characterize the concept of technology as a type of applied science, and to position it hierarchically under the scope of applied science.

Basic sciences (also called fundamental sciences) and applied science are sometimes confused with each other. While the basic sciences focus on facts, objects, or events and they seek to answer questions such as why, how, how often, and how much. The basic sciences help to develop theories. Applied science only centers on processes and actions and, unlike the basic sciences, is based on interventions in processes/actions with a new approach. For example, a new work safety approach can be implemented in an organization, or a new curriculum can be tested to improve educational practices. The success of the work safety approach or new curriculum is evaluated by comparing the results derived before the application (by looking at the past frequency of registered work accidents) and outcomes observed after the new safety application is implemented (by looking at the current frequency of work accidents). If a predetermined target standard exists and this standard is achieved, the new safety approach is considered successful, and it is recommended that the application be expanded. If success is not achieved, the focus is shifted to developing the approach. If the results did not make a difference compared to the previous one, the approach is abandoned. Thus, there is research in the application process. This research is generally based on comparing the results before and after the application is employed. Sometimes different applications can be tested at the same time. For example, an applied scientific approach in education can be based on comparing different teaching styles (Bickman and Rog, 2008; McMillan, 2015).

The goal of science and engineering is to build better mousetraps. The goal of nature is to build better mice.

Anonymous

Applied sciences focus on the use of theoretical knowledge obtained from basic research. The purpose of applied science is to develop new practices or to improve current ones in related areas. Applied science also helps to identify the places of use of theoretical knowledge (Bickman and Rog, 2008; McMillan, 2015).

All research types classified under quantitative and qualitative research approaches may have characteristics of the basic sciences or applied science.

The Relationship between Science, Technology, and Innovation

"Science, technology, and innovation are highly interdependent but also distinct" (Brooks, 1994: 477). **Technology** is the application (mostly industrial use) of scientific knowledge (Britannica, n.d.). This definition is, of course, not precise. It can even be said that it is a somewhat limited definition. McOmber (1999) offers two more definitions of technology apart from industrial context which are in the context of instrumentality and novelty. In the context of instrumentality, watermills, chariots, agricultural objects and even wheel and hand tools were also technological objects used by pre-industrial societies[10]. In the context of novelty, technology is everything "compelling reorganizations of social structures, values, and priorities" (McOmber, 1999: 144).

> **"Today's science is tomorrow's technology."**
> (Edward Teller & Allen Brown 1962)

Technology emerges with the conversion of scientific knowledge into technical knowledge. Science can contribute to technological advancements directly with new knowledge, such as nuclear and chemical knowledge, leading to nuclear and chemical technologies. Science can also contribute to technological advancements indirectly as a source of engineering design tools, techniques, analytical methods, and the development of human skills (Brooks, 1994).

As technological advancements proceed to higher levels, they also contribute to scientific progress. In short, there is an *interaction* between them. Due to this reciprocal relationship, higher technology enables the production of more sensitive, profound scientific knowledge (Brooks, 1994; Rescher, 2003), such as more precise measuring tools related to light and sound. Having more powerful computers today allows us to develop simulations that help us to better understand many natural phenomena. Advanced computer programs supported by high hardware technology increase our data processing capacity in many fields, especially in the social sciences. For example, thanks to the Internet, data collection is both easier and faster. For social research, it is possible to reach participants on a very large scale in a short amount of time. It is possible to explore outer space with advanced telescopes, and to more easily observe

[10] Further reading regarding technology and its historical context: "Basalla, G. (1988). The evolution of technology. Cambridge University Press."

micro-organisms with advanced microscopes. These examples can be further increased. In short, the relationship between science and technology is reciprocal.

"**Innovation** is the introduction of something new". More specifically, "innovation is a new idea, method, or device" (Merriam-Webster, n.d.). Innovation is not only related to developing a product. It extends to the refinement and advancement of a process, application, or mindset (Kahn, 2018). Whether it is linked to a product, process, application, or mindset, the support of scientific knowledge is important for innovation.

Positive Science and Normative Science

Keynes (1955) described **positive science** as the establishment of *uniformities*; **normative science** is about the *determination of ideals* and the *formulation of precepts* (Blaug, 1992; Bryman, 2007). Sometimes two qualities can be intertwined (Papineau, 2007).

Positive science is based on empirical investigation and is mostly *descriptive*. Normative (or regulative) science refers to norms and standards put forth by scientific studies. Some areas of science and disciplines are normative in nature. Normative scientific fields include spheres such as[11] *aesthetics, ethics, logic,* and *epistemology* (Bishop and Trout, 2016; Graham, 2014; Papineau, 2007). Occupational health and safety management is a normative discipline among applied sciences (Da Silva and Amaral, 2019). Some realms of science, such as economics, exhibit aspects of both positive and normative science (Galbacs, 2020; Hands, 2012). Some applied fields of science are also normative, such as engineering.

If we look from the viewpoint of research methodology, scientific research can be considered a process that entails normative standards (Blaug, 1992; Bryman, 2007). However, normative science and normative essentials of science are not the same thing. In scientific research, as in everyday life, some procedures may become norms over time. Norms in everyday life become our normal, and in science norms, they become science's normal. The accepted normal produces normative instructions.

In particular, applied research may aim to set standards in the realization of a business or activity. Research can introduce various principles and rules, along with the objective results of (an) application(s) developed to improve business (or activity) standards. In this respect, applied research can be normative.

---

[11] Some of these titles are considered partially normative –e.g. aesthetics-, while others are controversial –e.g. logic- in the world of philosophy of science.

Deterministic Science and Probabilistic Science

Determinism is based on the understanding that there are triggering causes behind all events (or actions). A person who defends this understanding is called a determinist, and thinks that the reasons behind all behaviors can be determined. For example, Sigmund Freud (1856–1939) has been termed a psychical determinist because he thought that human behavior reflected a series of unconscious impulses and motivations (Weathington, Cunningham, and Pittenger, 2012).

Determinism can also be defined as the view that everything that happens is necessitated by what has already gone before. Thus, another inference of the deterministic view is that nothing can happen differently from how it does (Strawson, 2005). At this point, we come across the *free will* problem of determinism. Determinism has been criticized for excluding *free will.* A view that accepts free will is based on the perspective that people's cognitive and mental states guide their behaviors. Supporters of free will argue that human behaviors cannot be understood through scientific methods since each person is unique (Weathington, Cunningham, and Pittenger, 2012).

An alternative view, called *compatibilism*, suggests that free will is compatible with determinism because freedom is essentially just a matter of not being constrained, or hindered, in certain ways when one acts or chooses (Strawson, 2005).

It is meaningful to examine the factors behind common behaviors, as people share some basic characteristics, and the deterministic stance suggests that observed behaviors can be explained by exploring the underlying causes. As a result, dealing with science requires both a deterministic view and a perspective on free will (Weathington, Cunningham, and Pittenger, 2012; List, 2014).

Deterministic science is mostly associated with the natural sciences. For example, Newtonian physics is an example of deterministic science. However, in light of contemporary developments, for example, it is controversial whether quantum physics (which has supplemented classical physics in the natural sciences) is deterministic (Gouinlock, 2005).

Probabilistic science is based on probabilistic computations, which are non-deterministic. Probabilistic calculations are rooted in assigning a probability to each possible outcome. However, probabilistic science is only sufficient when the outcomes are countable (Jones, 1990). Pearson (1911) believed that we have lived in a probabilistic universe. He considered causation to be a correlation and accepted correlation as an association[12]. Hence, his successors always sought laws of relations that could be statistically expressed. This perspective considers scientific cases in terms of probability. In other words, causal relationships can be reduced to probability calculations (Mulaik, 1986). For this reason, the probabilistic view of science is closely related to mathematics and statistics. For example, the Bell Theorem was initially confirmed mathematically and then experimentally. The theorem indicates quantum mechanics has a probabilistic nature (Wayne, Genoves, and Shimony, 2021).

In sum, it is deterministic science that helps us understand the world and the universe we live in. Inferences we arrive at via our perceptions, observations, or experiments are represented in deterministic science. On the other hand, when we arrive at the quantum world, a world of probabilities confronts us, in which we can only make science through probabilities. For example, The Nobel Physics Prize of 2022 has been awarded for experimentally confirming the Bell Theorem which indicates quantum mechanics has a probabilistic nature. Albert Einstein, Boris Podolsky, and Nathan Rosen had previously asserted a deterministic approach called the EPR paradox[13] to explain how quantum mechanics works. Alain Aspect, John F. Clauser, and Anton Zeilinger have been awarded The Nobel Physics Prize for the confirmation of John Stewart Bell's mathematical calculations through experiments. Their experiments also falsified the EPR paradox[14] , which contradicts the probabilistic approach of quantum mechanics.

CONCLUDING REMARKS

It is anticipated that readers know what science does, which activities are scientific, and what features are necessary for pursuing science after reading this chapter. Readers can also distinguish branches of science and classifications of scientific activities according to the nature of pursuing science.

---

[12] It is very important to understand causal relations. Furthermore, I should indicate that causality and correlation do not have the same meaning. Causality and its relation to correlation and its difference from correlation are examined in Chapter 3, under the title -*Causality.* However, as preliminary information, causality is a logical argument.

[13] Einstein, A., Podolsky, B., & Rosen, N. (1935). Can quantum-mechanical description of physical reality be considered complete?. Physical review, 47(10), 777.

[14] EPR paradox became identical to Einstein's saying –god does not play dice– which indicates probabilistic nature of playing dice. Einstein's use Word –God- is a metaphor here and indicates laws of nature. EPR consists of the first letter of the surnames of the authors.

SUMMARY OF CHAPTER 1

Science consists of activities to produce knowledge. Science uses systematic methodologies to produce accurate and reliable knowledge. Logic and philosophy help the mind process the collected data to make inferences. There is no place for miracles, mysteries, intuitive predictions, or faith in science. Questioning is one of several of the starting points of science. Curiosity or intention to solve a problem is also important to pursuing science.

A scientist asks questions, searches for data, and performs analyses. A scientist should be objective, skeptical, open-minded, and curious. Asking questions necessitates logical basis and theoretical knowledge, because the accurate way of asking questions is to ask the right question in the right way.

There are many *classifications of scientific fields* according to different institutions and for several purposes. The most fundamental and general distinctions of scientific fields are natural sciences and the social sciences and humanities. There are also some classifications of science according to the nature of pursuing science such as; *basic and applied sciences; positive and normative sciences; deterministic and probabilistic sciences.*

Chapter 2

PHILOSOPHY OF SCIENCE

Before starting to read the chapter

Philosophy is an extraordinary human endeavor in every aspect, and we know its historical background as far as the legacy left to us. If there were written texts lost in history, which we can assume there were, or if philosophy had a history dating back to before the invention of writing, we do not know. But it begs the question of who in history attempted to do such a thing or why it was needed. Perhaps philosophy, just like the invention of writing, may have arisen from the emergence of the public organization we call the state and its needs. Because we know that discussion of politics was very common in ancient philosophy. But man's preoccupation with thoughts may have made philosophy necessary as a result of the search for answers in an era when knowledge of nature was limited. This interesting human endeavor is closely related to science as well as politics. In this chapter, we will discuss philosophy in the context of its relationship with science.

This chapter contains four sub-sections, namely, the *philosophy of science, logic, epistemology, and scientific ethics.* This chapter aims to prepare its readers for scientific thought by presenting the philosophical, ethical, and logical foundations of science, especially after introducing science and explaining its activities and types.

After completing this chapter, readers recognize the classical and modern approaches to the *philosophy of science.* Furthermore, readers gain fundamental information about *logic* which helps them to use deductive and inductive reasoning through this chapter. Readers will also comprehend the verification requirements of the truth by getting to recognize *epistemology* and related approaches. Finally, readers are also anticipated to understand what the *ethical standards* of pursuing science are and how these standards should be implemented in different fields and various situations.

The chapter starts with ancient and modern approaches, and continues with contemporary approaches to the philosophy of science, which is grounded in the philosophical explanations of science. In this chapter, readers find out how developments in the philosophy of science and scientific progress go hand in hand.

The logic is then addressed because it is important for providing the basis for *reasoning.* Logic also provides the basis for a philosophy. In this chapter, logic is discussed on a basic framework that most of the readers are probably already familiar with from their high school education. On the other hand, some may not find it necessary to study logic when learning of scientific methods. However, even if you think that you do not need a basic understanding of logic, I believe that it will be useful to familiarize yourself with scientific reasoning methods such as deduction and induction, together with the basic notions of logic. Researchers may make reasoning errors when creating the research model or inferential errors when evaluating their findings, which is also discussed among the epistemological problems in this chapter.

Epistemology is discussed in detail via approaches of the philosophy of science in the context of verification. *Ethics* is one of the main topics of philosophy, and in this chapter; ethical problems are discussed in the context of pursuing scientific research activities. Ethical subjects are, at the same time, a subject of debate within fields of study since field-specific ethics guides field-specific standards of making science and applications. For this reason, it is important for readers that a general understanding of ethics should be followed by field-specific ethical developments.

PHILOSOPHY OF SCIENCE

Philosophy is defined as "the use of reason in understanding such things as the nature of the real world and existence, the use and limits of knowledge, and the principles of moral judgment" or, in other words, "the study of the nature of reality and existence, of what it is possible to know, and of right and wrong behavior, or a particular set of beliefs such as philosophy of Kant" by Cambridge Dictionary (n.d.). Those definitions of philosophy show us that we can benefit from philosophy while seeking the truth and establishing our relationship with knowledge. Because philosophy includes truth-related values, meanings, and the patterns of logic people perceive reality and experiences related to these perceptions.

Many philosophical studies have been conducted that examine knowledge and science from past to present in connection with advances in science. Although different sources have presented the development of the philosophy of science with various historical sections, it is possible, based on common points, to divide the development of the philosophy of science historically into four main periods. Those are ancient times (refers to the period between BC 2000 to BC 300), the scientific revolution[1] (refers to the period between the early

[1] It can easily be seen that there is a large periodic gap between ancient times and the scientific revolution. Of course, there were scientific developments during this period, in accordance with the focus of this chapter we will exclude this long period since there was no significant development in terms of the philosophy of science.

16th and late 17th century), modern philosophy (refers to the period between the 17th and early 20th century), and contemporary approaches. Figure 2.1 illustrates how the philosophy of science has evolved over time.

Figure 2.1. The Development of the Philosophy of Science

Ancient Greece (refers to the period between the 15th century BC to the 3rd century BC) is widely regarded as the birthplace of philosophy. Many of us are familiar with famous philosophers such as Socrates, Plato, and Aristotle. Ancient Greece is also recognized for its developments and inventions in the fields of science.

In terms of political science, it is the place where the concept of *democracy* was invented. There were also important developments in the field of medicine in Ancient Greece. The *Hippocratic Oath*, universally recognized as a commitment to human health, stands as a cornerstone of medical ethics. Similarly, the *Archimedes Screw* exemplifies ancient mechanical ingenuity. *Plato's Academy* played a pivotal role in the dissemination of science and philosophy. Meanwhile, the *Pythagorean Theorem*, commonly taught in geometry classes, is attributed to the Greek mathematician Pythagoras, who lived around the 6th century BC.

Of course, it should not follow from this that there were no significant developments in the rest of the world. Early traces of some developments can be found in different geographies, even going back much further than Ancient Greece. For example, early traces of the Pythagorean Theorem can also be found in the Babylonians as far back as around the 18th century BC (Hoyrup, 1999) and possibly in the Egyptians around the 19th century BC (Lumpkin, 1980).

Besides, we come across the endeavor for the classification of scientific activities in Ancient Greece. During this period, studies in *arithmetic, geometry, music, astronomy, logic, politics,* and *rhetoric* were common. In Ancient Greece, scientific methodology (e.g. Aristotle's studies on *induction* and *deduction*) was also studied.

With the scientific revolution, which covers the 16th and 17th centuries, experimental and evidence-based science methods emerged. This development is closely related to the philosophy of science because the philosophy of science guides the *scientific methodology* and *principles.*

Philosophy of science (as natural philosophy) developed with the epistemological studies of Galileo, Newton, and Kant, and advanced with the contributions of Bacon, Descartes, Leibniz, Maxwell, Popper, and many more scientists and philosophers (Maxwell, 2012; Worrall, 2005). The philosophy of science guides scientific *consistency* (e.g., logical suitability in terms of cause-effect relationships), *internal integrity* (in terms of modeling), and the *production of systematic outcomes* (regarding validity and reliability).

Science contains features that characterize it according to a classical understanding, and which have been shaped by the efforts of *logic, epistemology,* and the *philosophy of science.* The modern understanding of making science is a more flexible version of the classical understanding and considers conditional states. Classical understanding includes these assumptions regarding science:

- ⇨ is factual and based on observation
- ⇨ is rational, based on logic, and starts with a hypothesis (or question)
- ⇨ can be examined (through experiments or observations), such that there is an effort to support a hypothesis
- ⇨ is controllable –sometimes- (through tests/measurements)
- ⇨ can be generalizable (universality-under certain limitations)
- ⇨ is objective[2]
- ⇨ is critical; and
- ⇨ advances cumulatively (paradigm shifts can also occur)

Although today's understanding of science still carries some aspects of the classical understanding, it has a more flexible understanding, as we pointed out in the explanations. In accordance with the name of the book, we will discuss the content of the classical understanding of science presented above within the modern understanding throughout the book.

Many philosophical approaches arising from the past to the present tried to explain how scientific methodology and analysis should be conducted and what features accurate knowledge should possess in order to represent reality. In other words, approaches of the philosophy of science are concerned with the nature of *reality* and *knowledge* and how reality and knowledge can be achieved. These philosophical approaches, as scientific worldviews, discuss the nature of *reality* in the context of ontology (closely related to metaphysics) and *how to reach the knowledge of reality* in the context of methodology. Some

[2] Some disciplines may permit and include subjectivity partially such as *history* and *politics.* Even so, this freedom does not mean that the researcher/scientist is completely free from objectivity. All the studies researchers conduct in these areas must follow the objectivity principle as much as possible.

of those approaches are *reductionism, social constructionism, scientific realism, pragmatism, naturalism, empiricism, and post-positivism.*

Reductionism argues whether a particular subject can be reduced to another. In other words, reductionism is interested in features, concepts, and methods that can be used to produce conclusions from one scientific field to another (Brigandt and Love, 2023). For example, biology or chemistry can be reduced to atomic physics (Brigandt and Love (2023) point out "*biological properties supervene on physical properties*"). More specifically, "chemical properties like the solubility in water or valence have been reduced to the properties of molecules and atoms". Similarly, "laws of optics are reducible to principles of electromagnetic theory". Reductionism, at its most assertive claim, suggests that all sciences can be reduced ultimately to physics which is also known as the *unity of science* (Kim, 2005).

From a methodological perspective, reductionism recommends, for example, examining molecular or biochemical components in order to understand biological systems, meaning the lowest possible level of a system should be studied for a better understanding of upper systems (Brigandt and Love, 2023).

Reductionism also discusses the knowledge (as epistemic units) of a particular field of science that can be reduced to the knowledge of another field of science, which also means theories (as a set of knowledge) can be reduced to one another. For instance, *-is evolutionary biology reducible to molecular biology? -* (Brigandt and Love, 2023).

Constructivism, as a philosophy of science, is related to "*the meaning-making activity of the individual mind* (Patton, 2014: 201)" **Social constructivism** argues that social reality is subjective by its nature, and the learning process happens through social interaction (Fox, 2008). Culture is at the center of social constructivism. This is because culture and its social context are decisive in social functioning and production of knowledge. In other words, according to social constructivists, knowledge is a product of human understanding and is constructed in the context of culture (Jackson, Karp, Patrick, and Thrower, 2006).

According to social constructivists, in parallel with their view of knowledge, the reality about the world is also produced by individuals (Jackson et al., 2006). So, reality is not independent of actors (e.g., individuals, organizations) and emerges as collaborative conclusions of *concepts, values, beliefs, ethics*, and *norms* of the actors (Fox, 2008).

The social constructivist approach to scientific methodology requires researchers to treat the experience of each individual taking part in a particular social context in a case to be examined as unique and real. This may require using open-ended interviews and observation methods to measure individuals' unique experiences. Thus, the researcher can reveal "multiple realities" based on different perceptions (Patton, 2014: 201-202).

Another current understanding within the philosophy of science is scientific realism. "**Scientific realism** asserts that the objects of scientific knowledge exist independently of the minds or acts of scientists and that scientific theories are true of that objective (mind-independent) world" (Fine, 2005: 950). Scientific realism exhibits a dual characteristic. It claims the independent existence of certain entities, which is a metaphysical (specifically, an ontological) doctrine, and it also claims that we can know what individuals exist and that we can find out the truth of the theories or laws that govern them as an epistemological doctrine (Fine, 2005: 950).

Pragmatism approaches reality, assuming that "*a statement is true if it works.*" With this feature, the pragmatic approach, as a philosophy of science, focuses on the solution-providing nature of research or its practical benefit (Patton, 2014). Briefly, pragmatism appoints purposefulness at the center of scientific research studies. This approach prioritizes the utilities of scientific activities for the purpose of solving problems and improving the human experience (Shields, 1998).

Pragmatism is usually associated with mixed methods of scientific research practices. The pragmatist approach facilitates the design of research with the aim of purposefulness and supports the use of qualitative research for this goal and its inclusion in the research design as a part of mixed methods (Mitchell, 2018; Patton, 2014). The pragmatist approach is used in applied science studies such as those conducted in education and business administration.

Naturalism, as a philosophy of science, asserts that our universe functions on natural laws and forces (Barker and Kitcher, 2014). Pioneering philosophers (e.g., Isaac Newton in the 17th Century) who adopted this approach emerged with modern philosophy. Naturalist philosophers at the time started to seek answers to the question of the validity of scientific knowledge and norms for scientific understanding in their methodological debates. Naturalism aimed rightfully to free science from a religious understanding of the world and offered an understanding aimed at saving our efforts to understand nature by appealing to divine or other supernatural powers (Rouse, 2007: 62). Through its basic approach, naturalism made a significant contribution to the development of the philosophy of science.

Naturalism, in the context of the philosophy of science, has two dimensions, namely *ontological naturalism* (sometimes called metaphysical) and *methodological naturalism.* Ontological naturalism posits that the natural world is all that exists, and there are no supernatural or metaphysical entities or forces. This perspective asserts that everything that exists, including consciousness and human behavior, can ultimately be explained by natural phenomena and laws (Papineau, 2016; Rosenberg, 2012). Methodological naturalism is an approach in science that limits scientific inquiry to natural explanations, without invoking supernatural or metaphysical causes. Scientists operating under methodological naturalism seek to understand phenomena through empirical observation, experimentation, and the formulation of naturalistic theories (Coyne, 2015; Pigliucci and Boudry, 2013).

Although the naturalist movement, in the context of science, initially established its foundations on removing supernatural beings from the center of science, later criticisms suggested that naturalism did not exclude supernaturalism since its arguments assert that science does not have the basis to discuss supernatural subjects. Naturalist philosophy of science is today still criticized for holding its understanding of science as non-excluding supernatural entities, and underestimating philosophy's guidance to science (Papineau, 2007; Clarke, 2009), asserting that philosophy cannot restrict scientific methods (Rouse, 2007).

Another movement in the philosophy of science is **empiricism** which indicates that we can reach knowledge based on experience. Empiricism asserts that pure reason cannot produce useful and substantive knowledge of the world. Therefore, empiricists' model of science is based on experiments[3] and observations. According to empiricists, reason can only produce relations among our concepts (Ladyman, 2007; Patton, 2014; Schwandt, 2001).

Empiricists usually emphasize sense experience, excluding other types of experience such as *aesthetic, moral, religious,* and *abstract thinking.* According to empiricists, we gain knowledge through the stimulation of the five senses. Most critiques of empiricism related to its assumption of knowledge based on sense experience. Actually, all knowledge does not stem directly from sense experience (memories and previous thoughts can help in the production of new knowledge [Connolly, 2023]). Additionally, types of knowledge (e.g., mathematical knowledge), in effect, restrain the empiricist claim referring to experience (Alston, 2005: 221). Afterward, empiricism ushered in logical empiricism, which is closely related to logical positivism (Creath, 2023) and is discussed in detail in the following pages.

The classical understanding of science is best represented within positivism. The term *positivism* was coined in the 19th century by French philosopher

---

3 Schwandt (2001) indicates the difference between *empiricism* as the philosophy of science and *empirical research.* For example, all forms of qualitative inquiry are empirical research to the extent that they provide data of experience. It means that qualitative inquiry is an empirical research; however, it does not mean that it rests on the philosophy of empiricism (Patton, 2014).

Auguste Comte (1798–1857) in his early works (1820-1829). It is similar to the deterministic approach of the natural sciences and continued to have an influence in the early 20th century (Bourdeau, 2022; Glesne, 2011; Kincaid, 2005).

In essence, positivism emerged as an empiricist movement. Kincaid (2005) mentions that "the basic positivist understanding suggests that philosophy should be scientific." According to the positivist lens, "metaphysical speculations[4] are meaningless." **Positivism** asserts that "there is a universal and a priori scientific method that a main function of philosophy is to analyze that method, that this basic scientific method is the same in both the natural and social sciences." (Kincaid, 2005: 826).

In many fields, both natural and social scientists use positivist methods and concepts (e.g., validity, reliability, objectivity, generalizability) in their research (Fox, 2008; Patton, 2014). In this respect, positivism attributes importance to quantitative data and analysis and is skeptical of qualitative methods such as interviews (Fox, 2008; Kincaid, 2005).

A positivist researcher has an ontological view that there is only one reality presented to humans, which can be measured and has a particular degree of certainty (Glesne, 2011). Positivist philosophy is also founded on the idea that various sciences can be reduced to physics. All these views have received criticism.

Positivism is criticized for its claim to one reality. Specifically, it receives critiques in the context of the nature of the social world. This critique stems from the view that social order emerges based on condition-dependence. Furthermore, positivism is criticized for trying to make context-free generalizations. In reality, scientific research can be established in a particular context of interest (Fox, 2008).

Positivism also receives criticism regarding its attitude against qualitative methods. Qualitative methods produce fruitful knowledge for many areas of social sciences (Fox, 2008) and can help to develop theories (Fraenkel, Wallen, and Hyun, 2011).

---

[4] Metaphysics is one of the branches of philosophy concerned with the nature of reality and can also be attributed to physics (Fine, 2012). In ancient philosophy, metaphysics was called the *first philosophy* by Aristotle. He treated metaphysics as an investigation of the *first causes* (Loux and Crisp, 2017; Bechtel and Hamilton, 2007). With this feature, metaphysics has been closely related to ontology and epistemology (Callender, 2011). The positivist view suggests that objects to the supernatural claims of metaphysics since supernatural entities and realms cannot be falsifiable and cannot be theorized. In other words, positivists critique metaphysics for its commitment to the mind-independence of the unobservable world (Ladyman, 2012).

Positivism transformed over time, and in the twentieth century **logical positivism** (later known as logical empiricism) appeared as another movement of empiricist philosophers. Logical positivism basically retained the fundamental principles of positivism (Ladyman, 2007). According to logical empiricists, *reality* exists regardless of whether we have any sense or consciousness of it. This view is also shared by other philosophical movements called *objectivism* (Patton, 2014) and *scientific realism* (Fine, 2005).

Logical empiricists' (or logical positivist) discussions of knowledge are based on three components. For them, knowledge about the world depends upon; *meaning, concepts,* and *verification* (Westphal, 2017). Logical empiricists defended the claim that knowledge and meaningful propositions can only be verified empirically and analytically. This principle is also known as the *verifiability criterion of meaning* (Ladyman, 2007). On the other hand, even among positivists, some were skeptical about *meaning* and *concepts* that are assumed to be coherent with the reality of scientific practice (Salmon and Glymour, 1999).

Post-positivism expresses a more flexible form of positivist understanding according to some scientists, while according to another group of scientists post-positivism is everything except how positivism was initially conceived. According to the second group, post-positivism is a way of thinking that includes all other paradigms except the form in which positivism was first introduced (Glesne, 2011). Post-positivists accept that there is not only one reality. In other words, the post-positivist view acknowledges that reality varies under different circumstances (Fox, 2008). Post-positivism considers all methods to be imperfect and recommends multiple methods be utilized together to create and test theories.

An overview of research paradigms gained momentum with Thomas Kuhn's book *The Structure of Scientific Revolutions,* published in 1962. His book undermined the dominance of logical positivism (also logical empiricism) in the philosophy of natural and social sciences (Rosenberg, 2008). Prior to the publication of this book, Western scientists tended to believe that research was self-centered, and that a person's knowledge base would gradually increase until she/he understood how the world works.

Kuhn's paradigm maintains that science does not increase gradually toward truth. According to Kuhn (1962), scientific history from Aristotle to Einstein can be characterized by revolutionary changes (called paradigm shifts) instead of overall improvement. In this respect, the history of science is the succession of very general theories called *paradigms* by Kuhn. According to Kuhn, Copernicus' Heliocentric Model, Newton's Laws of Motion, Darwin's Study of the Origin of Species, and Einstein's General Relativity are all samples of paradigm shifts in the history of science.

Kuhn used the term *paradigm* broadly. The term refers to *research communities, concepts, social,* and *material components* that serve a paradigm shift. This paradigm change makes scientific activities coherent and stable by identifying and framing those activities (Ankeny and Leonelli, 2016; Kuhn, 1962).

Kuhn claimed that organized communities of specialists are the central forces in scientific development (Kuhn, 1962). Kuhn also argued that knowledge is limited to what is *reasonably extracted from theory,* has *measurability,* and is *empirically repeatable* (Glesne, 2011; Patton, 2014: Rosenberg, 2008). Kuhn has made significant contributions to the realm of scientific endeavors by enhancing comprehension regarding the interplay between individual contributions and communal practices and standards. He also indicated the importance of the association between research practices and policy, funding, public relations, marketing, and institutions (Ankeny and Leonelli, 2016).

Contemporary philosophy of science increasingly focuses more on the contextual use of research methods in the various sciences rather than routine practices of pursuing science (Haig, 2018) and examines the relationship of results obtained in light of *measurements, observations,* and *experiments* in terms of scientific knowledge. In other words, the modern philosophy of science sees theories as relatively representative of truth and evidence as a means of verifying theories that are put forward.

According to the modern philosophy of science, it is important to note how close scientific knowledge is to reality. We want to know that science's explanation of an entity (fact, object, or event) truly represents the underlying reality. The absoluteness of this state of being true, however, will be debatable. Perhaps it would be more accurate to say that it is *probably true* or *approximately true,* even *probably approximately true* (Chakravartty, 2017), as asserted by scientific realism. For example, if we are talking about the world of subatomic elements, the measurements made, and the findings obtained will likely be an approximate representative of the truth. This standpoint calls for using adjectives such as *adequate* and *reasonable* when describing scientific knowledge and theories.

Table 2.1 presents the basic *ontological, epistemological,* and *methodological* views of the approaches explained above in the context of the philosophy of science. The ontological dimension is, in some sources, referred to as the metaphysical dimension.

Table 2.1. Ontological, epistemological, and methodological views of various approaches in the philosophy of science

Philosophy of science	Ontological/metaphysical claim	Epistemological claim	Methodological claim
Reductionism	The properties and processes of a system supervene upon another system. This means that higher-level phenomena are determined by and dependent on the properties of lower-level entities	Relations can be established between different scientific fields by carrying knowledge and theories from one to another	The lowest possible level of a system should be studied for a better understanding of upper systems
Social constructivism	Social reality is constructed by individuals and groups and is subjective by its nature	Knowledge production depends on culture and its social context	The experience of each individual is unique and can be understood in individual's particular social context
Scientific realism	A mind-independent reality exists. This means that the world and its entities exist independently of our thoughts, perceptions, and theories about them	Scientific theories partially represent reality about our world	Improvement of the theoretical explanations is crucial
Pragmatism	The nature of experience is essential. Reality is shaped by practical experiences and actions	A statement is true if it is beneficial	A research study should aim to solve problems or increase benefits for humans
Naturalism	The universe functions on natural laws and forces	Theories are limited rationally	Philosophy should not restrict scientific methods but should limit scientific inquiry to natural explanations
Empiricism	Reality exists independently of our perceptions, but our knowledge of it is derived from sensory experience	Foundations of knowledge should be as durable as they do not require further justification	Knowledge can be produced by experiments and observations/ We gain knowledge through our senses
Post-positivism	Reality varies under different circumstances	Knowledge is relative rather than absolute	Multiple-method use is essential for creating and testing theories

Source: Brigandt and Love, 2023; Fine, 2005; Fox, 2008; Jackson, Karp, Patrick, and Thrower, 2006; Kincaid, 2005; Papineau, 2016; Rosenberg, 2012; Shields, 1998.

LOGIC

Logic is the study or act of proper reasoning (Vander Nat, 2010). This definition refers to the feature of logic as a way of thinking. Logic is at the core of our reasoning skills and is based on laws and principles. Besides, it is a way of generating new information based on available data. Logic also considers similarities and divergences of subjects and exceptional cases in the process of reasoning (Smith et al., 1992).

There are many other definitions of logic as well (Bach, 2006; Jago, 2007; Vander Nat, 2010). Logic is:

⇨ the knowledge of the rules and methods of proper reasoning

⇨ the science of the laws of reasoning

⇨ the study of formal conditions of linguistic expressions

⇨ the theory of correct propositional forms; that is, precise statement patterns

The foundation and development of logic as a discipline has been a part of the history of philosophy since Aristotle (Vander Nat, 2010), who developed the first form of logic (Evans, 2005). Classical logic forms the basis of rational thought. However, studies have shown that most people do not reason abstractly. On the other hand, some studies suggest that people retain their ties with their personal beliefs while achieving inference through reason (Evans and Over, 2013). Personal beliefs may cause objectivity problems in scientific studies.

Logic deals with the formal relations of reasoning between thoughts, regardless of how a thought forms (the proposition) and its content (conceptual elements) in general. As a scientific field, logic examines the laws of thinking. In other words, it scrutinizes how a judgment is made and how a correct judgment should be made. Thus, logic is concerned with the form of reasoning and studies these forms in terms of validity (Özlem, 1994).

The system of logic is generally considered bivalent; that is, statements are only *true* or *false* (Vander Nat, 2010). If these definitions are taken into account, then logic is not concerned with the content of a proposition. In other words, whether a proposition is true or false is not a matter of logic. Whether the content of a proposition is correct or incorrect is a matter of knowledge. In other words, the truth or falsity of a proposition is the subject of the relevant discipline and epistemology. For example, a physicist produces information (by *induction*) by repeatedly testing (via experiments) the proposition that "pure water boils at 100 °C." The scientist confirms this information by testing it many times; that is, knowledge is generated via science (in this example, physics). Generating results through induction is a process of logic. Therefore, we always produce results by applying logic in engendering scientific knowledge. The reasoning here is as follows (Özlem, 1994):

→ All my experiments show that pure water boils at 100 °C at sea level.

→ Thus, pure water always boils at 100 °C at sea level.

Logic deals with the principles and laws of reasoning. Logical principles and laws can be used to generate knowledge. We emphasize that the *principles, laws,* and *methods of thinking* are the subject of logic. Consequently, knowing the truth cannot be independent from logic, even if its contribution is indirect (Fraenkel, Wallen, and Hyun, 2011; Nickerson, 2011; Özlem, 1994). Accordingly, logic proposes four principles (Özlem, 1994):

✓ The principle of identity;

✓ The law of non-contradiction;

✓ The law of the excluded middle (or third state); and

✓ The principle of sufficient reason.

The **principle of identity** is the most basic principle that allows us to contemplate an entity's state of being. In other words, everything is itself, and no other thing. This thought naturally points to the state of being as nothing but an entity being itself in the next step, at which point the principle of non-contradiction arises. The **principle of non-contradiction** means ontologically that "something cannot be both itself and something else at the same time" (Marvin, 2011). For example, "all humans are mortal, so any human cannot be immortal." There is a nuance in this principle that works well. For example, if there is a valid premise indicating a contingency, then the proposition is not necessarily true, even if there is very little possibility for an alternative truth (Lewis, Langford, and Lamprecht, 1959).

The principle of the excluded middle means ontologically that "something is either itself or something else." The principle of the excluded middle can also be formulated as "something either exists or does not exist; there cannot be a third state" (Jevons, 1893).

The **principle of sufficient reason (PSR)** means that for everything existing, there is a sufficient reason. More clearly, everything must have a *reason, cause,* or *ground* (Lu-Adler, 2021). This principle concerns why something is what it is, and is related to cause-effect relationships (Marvin, 2011). Shortly, this principle claims everything has an explanation (Amijee, 2021).

In contemporary philosophy, some authors (e.g., Peter van Inwagen) suggest that the principle of sufficient reason collapses all modal distinctions. Opponents of this principle indicate contingent conditions (Lu-Adler, 2021; Tomaszewski, 2016). Proponents of PSR (e.g., Della Rocca) assert that if all existing facts demand an explanation, the PSR is true. Contemporary debates on PSR move around *necessitation*[5] and *grounding* (Amijee, 2021).

Another important term regarding logic is premise. A **premise** is a proposition that functions as evidence in proving the correctness of a result through inferencing (Vander Nat, 2010); it is put forward through reasoning. Let us look at an example of a syllogism via deductive reasoning (Fraenkel, Wallen, and Hyun, 2011):

All humans are mortal.
Sally is a human.
So Sally is mortal.

The first statement is the *major premise*. For this, it is necessary to make a generalization both about mortality and individuals. The *minor premise* is established after the major premise appears. At the last stage of logical reasoning, the *conclusion (inference)* is produced. Our judgment about Sally is based on information from our senses. We no longer need data from our senses to draw a conclusion. If the first two statements are true, the third one will be logically correct (Fraenkel, Wallen, and Hyun, 2011). Let us examine two other examples below (through deductive reasoning):

Example 1

Major premise: All living creatures are mortal.	**Deductive reasoning**
Minor premise: All birds are living creatures.	
Conclusion (inference): All birds are mortal.	

Example 2

All plants are alive. (Premise 1: Major premise)	**Deductive reasoning**
Daisy is a plant. (Premise 2: Minor premise)	
So the daisy is alive. (Conclusion/inference)	

Inductive reasoning, like deductive reasoning, needs premises to make conclusion. Let us examine two examples of inductive reasoning below:

Example 1

Premise 1: Every rose I water has shown signs of mortality after a period of time.	**Inductive reasoning**
Premise 2: Every rose I water is mortal.	
Conclusion: It is likely that plants, including roses, are mortal.	

---

[5] Necessitation is the principle that *a* grounds *b*, then necessarily if *a* obtained then *b* obtained (Amijee, 2021).

Example 2

Premise 1: Every bird we have observed has been a living creature.	
Premise 2: Every bird we have observed has been mortal.	**Inductive reasoning**
Conclusion: It is likely that all living creatures, including birds, are mortal.	

A general conclusion cannot be drawn based on a single observation for inductive inference. Observations need to be repeated.

The components of logic, from simple to complex, are *terms (concepts), propositions,* and *inferences.* The deductive process consists of these items. For this reason, I discuss the topics of logic from classical logic to the present under the headings of *terms (concepts) logic, propositional logic,* and *inferential logic* (Özlem, 1994). Modern logic is a branch of both mathematics and philosophy. A logical process begins with the relationship between a natural language and a formal language. Linguistic propositions (sentences) and mathematical formulas require the use of logic, and both represent the transition of actual features of certain fragments of a natural language into a formal language.

According to some logicians, the basic element of logic is a *concept*; that is, the smallest unit of thought. A logical concept can be anything imaginable. If you pay attention here, the concept (just like knowledge) is formed in the mind; thus a concept is a mental design. Some philosophers state that a concept is represented in terms because it is not independent of language. We truly need to express every concept as an idea (i.e., a representation expressed in language is called a term). Concepts are represented by terms in language. Terms are linguistic and phonetic symbols that refer to particular concepts. For example, when we say *cat,* it reminds us of the concept, but the term *cat* is not the concept itself. In other words, there is no identity linking the concept and the object in either existence or reality, since we acquire information about an entity through the senses. Thus, there can always be a difference between reality and appearance (Özlem, 1994).

It is vital for concepts to be *clear* and *distinct.*[6] This is logically and epistemologically important. The fact that concepts are clear and distinct has an epistemological aspect, as it is related to our knowledge of concepts. To be clear about a concept requires distinguishing that thing from other things. In

[6] Definitions of clear and distinct concepts were developed by Descartes along with the modern age of philosophy. If a concept is both clear and distinct, according to Descartes, it is an *evident concept.*

other words, something is itself and nothing else, which points to the two logical principles of identity and non-contradiction (Özlem, 1994).

The distinctness of a concept indicates features that enable it to be differentiated from other concepts. In short, if we look at it from a logical angle, the clarity of a concept denotes its scope, and its distinctiveness reveals its content (features). Concepts whose properties cannot be determined are *confused concepts* (Özlem, 1994).

Logic is concerned with concepts in terms of *singularity* or *generality* (Jevons, 1893). Concepts can be either *contrasting* or *contradictory*, depending on their ties to each other; these two are not the same. The fact that one concept contrasts with another does not mean it is contradictory. For example, if something is white, it is not black. Black and white are opposite but not contradictory,[7] since something does not have to be white if it is not black, or black if it is not white. For this reason, contrasting concepts always have middle concepts. For instance, there is a concept like gray between two contrasting concepts, such as black and white. However, there is no middle concept for contradictory concepts. For example, the contradiction of white is that of non-white (please note, it is not black); there is no middle concept between white and non-white, which shows that the principles of identity, non-contradiction,[8] and the non-third state in logic come into play here (Jevons, 1893).

Black and white	⇒	contrasting
Black and non-black	⇒	contradictory
Water is warm and cold	⇒	contrasting
Water is warm and not warm	⇒	contradictory

There are also concepts that are self-contrary and self-contradictory. Self-contradictory concepts are not valid either logically or in terms of the laws of nature. For example, "hexagonal circle" is a self-contradictory, invalid proposition in both logic and the laws of nature (Özlem, 1994).

A **proposition** is a sentence that states a correct or incorrect judgment. Each proposition is a linguistic construct that consists of at least two terms: the subject and the predicate (Jevons, 1893). Let us look at an example (Exner and Rosskopf, 2011):

---

[7] Hence, there is no obstacle for them to be in the same class. For example, black and white are included in the color class in terms of paintings. However, in the physical sciences, they are categorically excluded from the color class.

[8] Each proposition obeys the principle of non-contradiction in itself. For example, if we apply it to the first proposition, the contradictory proposition "A is not white" is rejected in terms of the proposition "A is white" because the two propositions contradict each other, as explained above.

Proposition 1 (premise 1): All humans (term) are mortal (term).
Proposition 2 (premise 2): Socrates is a human being.
Proposition as inference: So Socrates is mortal.

Propositions are divided into three branches in terms of relations: *categorical propositions* (*syllogisms*), *hypothetical* (*conditional*) *propositions*, and *distinctive propositions*. For example, the proposition "[a] human is a living being" is a categorical proposition. A hypothetical proposition is based on a hypothesis; that is, it expresses the interdependence (conditional aspect) of two concepts consisting of a major premise, a minor premise, and a conclusion. For instance, "if he took the 10:00 bus, he will be here on time. He took the 10:00 bus. Then he will be here on time" (Huby, 2004; Nickerson, 2011).

It is the job of the sciences to produce correct propositions in terms of content. Activities of knowledge acquisition require the use of logical laws and principles. As a result, knowledge generation activities—and therefore knowing—cannot be independent of logic. In other words, logic is a formal, instructive field on which truth-knowing (scientific) activities are necessarily based. Logic contributes indirectly to reaching the truth or knowledge (Vander Nat, 2010).

On the other hand, the use of logic is not as simple as I defined and exemplified above. Sometimes, philosophers and scientists come across challenging cases. Logical reasoning (for example, a hypothetical proposition based on two other propositions known as premises) of things might be meaningless if we do not select which features of those things are to count (Morreau, 2009). Besides, our minds and rationality have limitations. More specifically human mind has not been equipped to work by the rules of probability and through frequency formats (Crupi et al., 2008; Hertwig and Gigerenzer, 1999). In other words, our minds are inclined to get into a logical fallacy. This tendency sometimes stems from cognitive biases such as the *bandwagon effect*, which demonstrates a tendency to adopt certain beliefs, behaviors, and attitudes because many people adopt them (Bennett, 2012).

Philosophy acts on logic. Thus, the input of philosophy is logic, which has a close relationship with different types of mental functioning, such as *analytical thinking*, *dialectics*, *metaphor*, *analogical reasoning*, *analysis*, and *synthesis*. I discuss these concepts in Chapter 4.

Deduction, Induction, and Abduction

Deduction and induction are logical inferencing methods used to determine a general characteristic, and are based on known examples or observations (Rothchild 2006). Both deduction and induction have long been employed in scientific research and logical analysis/inference (through reasoning). They are generally weak in terms of producing accurate results. However, their joint use (via synthesis) supports the development of better scientific methods and the production of more precise outcomes (Miller, Strang, and Miller, 2010).

Deduction involves reaching a reasoned conclusion (logically) from known facts. In other words, in the process of deduction, a conclusion is reached by considering general facts. Deduction is based on theory. Because of this, a hypothesis derived from deduction is also known as the theory's hypothesis (Åsvoll, 2014). For example, high-performing employees are extremely competent at their jobs. For example, "*If John is performing well, then he is very competent at his job*" (Sekaran, 2003: 27). A hypothesis is deduced from a theory in the social sciences. Therefore, a research study in the social sciences is initiated via a related theory. As a result, each research study in the social sciences tests a related theory. The point to be noted here is that a deduction begins with the expression of *premises*, which are assumed to be real. If a premise is correct, an estimate is made about what else might be correct. The accuracy of a premise, which is the starting point, is important in order to produce correct results from deductive inferencing. Otherwise, a wrong conclusion can be reached because deductive reasoning is very prone to error (Lewis-Beck, Bryman, and Liao, 2004; Miller, Strang, and Miller, 2010; Weathington, Cunningham, and Pittenger, 2012).

Figure 2.2. The process of deduction[9]

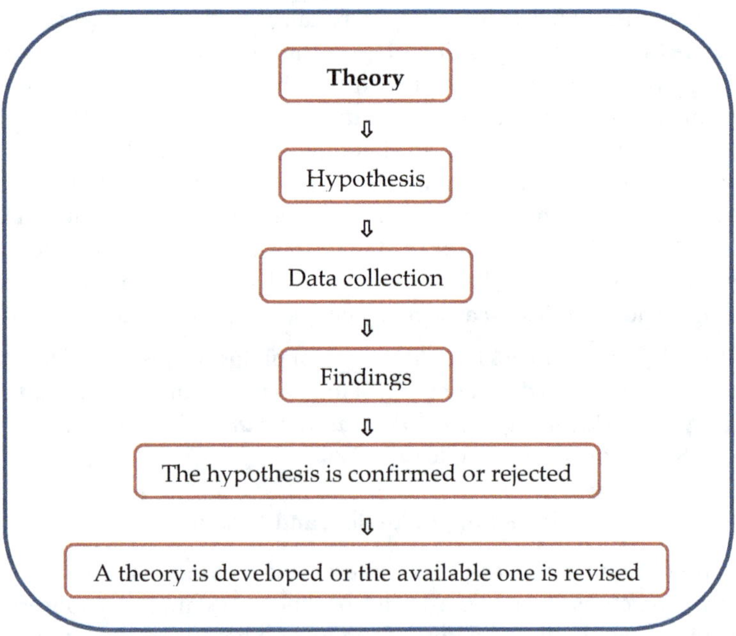

[9] Based on: Miller, Strang, and Miller 2010: 3 (quoted from Bryman, 2004)

Since deduction is a logical method, it requires following the principles of logic. However, it may cause false inferences to be made since systematic errors, such as biases, cause deviations from the principles of logic (Evans, 2005). The process of deduction is shown in Figure 2.2.

Let us consider another example, "*All living things die sooner or later. Trees are also living, so trees die too.*" The reasoning process in these two examples involves identifying a given feature by using a general judgment. The implications here are at two different levels (John is an individual; trees are a living class). The opposite of this inference technique can also be done. In this case, we will encounter the concept of *induction.*

Induction is the process by which scientists decide whether a theory is true based on various observations or experiments (Papineau, 1996). Induction is also a method of exploring general rules and principles rooted in specific facts and examples, according to the Oxford Dictionary (n.d.)[10]. According to another definition, induction is a method of reasoning (generalization or assumption) based on observed facts (i.e., based on available data; Oxford Dictionary, n.d.; Rothchild, 2006; Sekaran, 2003: 27). Using available data logically means producing conclusions derived from data. Induction allows one to determine which theory or theories can explain data. Induction does not prove that a theory is correct; it only allows us to identify the most suitable theory/theories for available data (Miller, Strang, and Miller, 2010). Let us take a look at the example below:[11]

→ With the apple falling to the ground, **observation** takes place.

→ The **hypothesis** is created by formulating the **thesis** (assumption/prediction) and is formulated as "*Objects fall on the ground when released.*"

→ The **experiment** is done. Objects are repeatedly dropped on the ground.

→ If the result *confirms* the theory, then the theory is supported; if the result *falsifies* the theory, then the theory needs to be changed.

[10] We should point out that, as Grove and Gray (2018) indicate, the theoretical definition of a concept and its dictionary definition may not be the same. "A conceptual definition is more comprehensive than a denotative (or dictionary) definition." (Grove and Gray, 2018: 192).

[11] Note that there is an attempt (induction) to reach a generalization here. The law of gravity is a natural law inferred via induction. Weather forecasts and predicting economic growth are more complex inferences made by combining the diverse body of evidence through induction (Sprenger, 2016).

Thus, a hypothesis is formed based on a single observation; the hypothesis can be proven through repeated experiments, and a generalization (induction) is achieved. The process of induction is shown in Figure 2.3.

Figure 2.3. The process of induction[12]

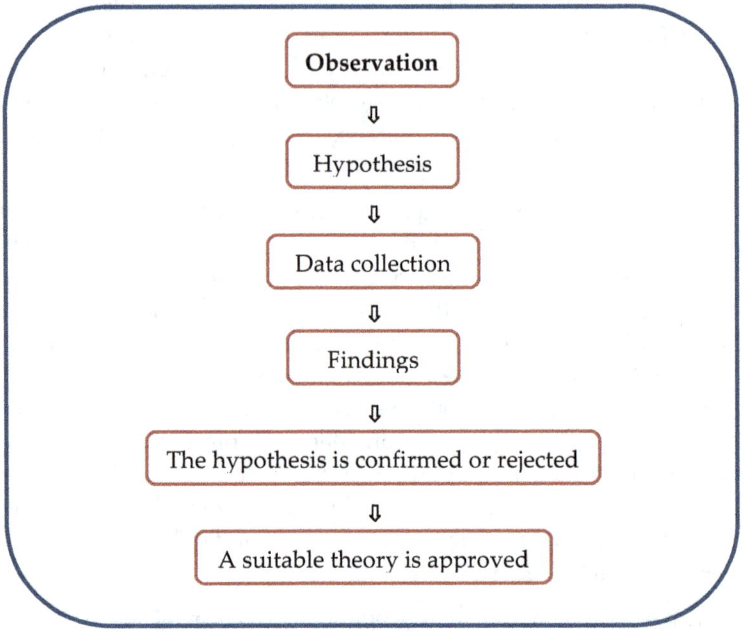

Put simply, induction allows us to make a generalization, while deduction allows us to make a reduction. If theories are constantly found to hold true, they become **scientific law**. That is why we call the *Law of Gravity* the *Law of Gravity*.

The basic idea behind reasoning through induction is to learn from experience. Experiences allow us to recognize distinctive features (characteristics and patterns) and resemblances (Lawson, 2005). Charles Darwin, for example, came across many finch birds in the Galapagos Islands and realized that the shapes of the birds' beaks were different on each island. After examining the beaks, he concluded that each shape was suitable for the environmental conditions on each given island. Based on this, Darwin developed a theory. He inferred that in the past, a single type of finch had come to the islands, but over time, each population of finches evolved in a different direction depending on the island's environmental conditions. Thus, the most important theory in biological science was developed through inductive reasoning (Lawson, 2005).

[12] Based on: Lawson, 2005; Miller, Strang, and Miller, 2010

The path followed here is: (1) data collection; (2) analysis of the collected data; and (3) theory development.

Let us look at another example. Based on the observation that each crow we encounter is black, the general judgment of "*all crows are black*" can be reached. Keep in mind we'll come back to this example.

Confirmation (or justification) or falsification of a theory by induction has long been used. Confirmation can be conducted in two approaches: qualitative or quantitative. While qualitative confirmation tries to illustrate reality through the "grammar" of a concept, quantitative confirmation tries to illustrate reality through probability (Sprenger, 2016).

Johannes Kepler (1571–1630) theorized that the Earth and other planets move around the sun in elliptical orbits. Observations based on this theory made it possible to predict the movements of the planets. Thus, as each observation made confirmed the predictions based on this theory, it was deemed to have proven itself (DeWitt, 2010). Likewise, Newton developed the *Universal Law of Gravitation* (by advancing Kepler's laws) and used the induction method (De Pierris, 2006).

However, although it has provided confirmatory evidence for over 200 years, it was later discovered that Kepler's theory was not entirely correct. This means that inferencing (and theory verification) by induction may not give us an absolute truth, and shows that observations supporting a theory should be supported by different measurements and tests. Namely, as in Kepler's theory, scientific theories from the past can turn out to be wrong by new pieces of evidence (DeWitt, 2010). For this reason, some philosophers suggest probability as a tool for confirmation. Probability, as a degree of confirmation, is especially useful during the reasoning process under uncertain conditions (Sprenger, 2016).

Another term related to logical inference is abduction. **Abduction** is essentially the first stage of any scientific inquiry and of all inference processes. In other words, abduction is the activity of looking for potential explanatory patterns within the facts of a phenomenon. It is useful to eliminate irrelevant explanations. For example, when you come across a bag of apples on the table in the living room in the morning, you start looking for possibilities for that occurrence. You consider the possibilities to be a burglar, aliens, or a new neighbor. You seem to have three possibilities. But when you looked around and saw nothing is missing, you eliminated the burglar possibility. If you do not believe in aliens you also eliminate that option. Now you have only one possibility. When you discovered that your balcony door is open and you remember the only person with access to your balcony door is your neighbor, you now have the best explanatory hypothesis. That process is called abduction and the presented occurrence was an example of induction, namely the abductive hypotheses seek theory (Asvoll, 2014). We want to make *inference to the best explanation,* which relies on the existence or quality of an explanation (Lombrozo, 2012). Inference to the best explanation is an approach designed to give a partial account of many inductive inferences. Abduction, particularly, is one version of inference to the best explanation approach. Many inferences for ontological phenomena are naturally described this way. For example, Darwin inferred the hypothesis of natural selection because it would provide the best explanation of the evidence he found. Inference to the best explanation can also be seen as an extension of the idea of `self-evidencing' explanations, which exhibit a curious circularity, but this circularity is well-tempered (Lipton, 2000).

Validity and *soundness* are also sought when making logical inferences. Tony Roy provides a good example in terms of logical validity and soundness. Let us take a look at it. Logical validity and soundness correspond to different ways an argument can go wrong. Consider the following two arguments:

	Only citizens can vote		All citizens can vote
(A)	Hannah is a citizen	(B)	Hannah is a citizen
	Hannah can vote		Hannah can vote

> The line divides premises from conclusion, indicating that the premises are supposed to support the conclusion. Thus these are arguments. But these arguments go wrong in different ways. The premises of argument (A) are true; as a matter of fact, only citizens can vote, and Hannah (my daughter) is a citizen. But she cannot vote; she is not old enough. So the conclusion is false. Thus, in argument (A), the relation between the premises and the conclusion is defective. Even though the premises are true, there is no guarantee that the conclusion is true as well. We will say that this argument is logically invalid. In contrast, argument (B) is logically valid. If its premises were true, the conclusion would be true as well. So the relation between the premises and conclusion is not defective. The problem with this argument is that the premises are not true — not all citizens can vote. So argument (B) is defective, but in a different way. We will say that it is logically unsound."

Logical validity is a fundamental standard for valid conclusions. But we should point out that there is a distinction between logical validity, which is discussed in this chapter and the philosophy of science concept of validity (See Chapter 7).

Some scientists and thinkers have suggested that induction has no place in science. For example, Karl Popper, a famous figure in the philosophy of science, rejected induction as a form of logic in scientific practice. Similarly, Nobel laureate biologist Peter Medawar implied that induction has no place in science. When Karl Popper rejected induction as a method, he gave the example of the crow (which is why I give the same example) and stated that induction would be invalid if a single white crow were encountered. While crows tend to be black, the blackness of all known crows does not prove that there are no white crows. It is ridiculous to look at the matter in this way, but the key thing to note is that most crows are still black, even if all crows are not black (Rothchild, 2006). As a result, both inductive and deductive reasoning can be used in scientific activities (McMillan, 2015; Rothchild, 2006). On the other hand, the fragility of the confirmatory nature of the inductive method should be known (Sprenger, 2016).

EPISTEMOLOGY

Epistemology is a special field that deals with knowledge, and scientific methods and is one of the main areas of philosophy. Epistemology examines what we know as well as how we know what we know. In this context, epistemology deals with the concept (nature) of knowledge, the sources of knowledge, and the standards of valid, reliable knowledge, and methods used by sciences. Epistemology does not accept chance predictions or beliefs derived from a desired idea as knowledge (Grayling, 2002; Klein, 2005). An epistemological view of knowledge requires us to distinguish between knowledge and opinion, or between good reasoning and poor reasoning. Moreover, epistemology also concerns the possibility of nature and the limits of human intellectual capability (Cruz, 2006).

Epistemology is a normative field and also has principles as Lumer (2000: 408) mentioned, “The deductive epistemological principle: 'A proposition is true if it is *logically* implied by true propositions'; or the genesis of knowledge principle: 'A proposition is true if it has been *verified* correctly'; or the interpretative epistemological principle: 'A proposition is true if it is part of the only possible *explanation* of a known fact' etc.”

Since epistemology is concerned with truth, we also need to understand the means of accessing truth. Because epistemological verification entails only those statements that are true of logic and those statements that are empirically verifiable. In other words, we can reach the truth by reasoning through an empirical fashion. We can appeal to Leibniz (1646-1716) to comprehend the idea stated above. Leibniz made a distinction between *truths of reasoning* and *truths of fact*. Leibniz stated that if the analysis is limited, the ultimate truth can be reached through reasoning and expressed this as the truths of reasoning. If the analysis is unlimited, he said that we cannot reach the ultimate truth by reasoning, we can only reach it through experience, and he expressed this as the truths of fact (Look, 2020).

Epistemologically accurate knowledge necessitates verification (or justification) (Cruz, 2006). There is no consensus on the concepts of verification by scientific methods. But axioms and premises can be considered the first step of verification based on logic. In fact, it can be said that before experimental verification logic-based axioms and premises comes first. Because we establish hypotheses on postulates based on logic (Barker and Kitcher, 2014). Verification can be accomplished through *personal experience, scientific investigation,* or *mathematical calculation* (Rescher, 2003).

There are various philosophical approaches to verification in epistemology. One of them is *scientific pragmatism,* which accepts that single or multiple realities can exist considers situations/conditions that can affect verification and sees evidence as insufficient for verification (Fantl and McGrath, 2002; Kaushik and Walsh, 2019).

Philosophical naturalism also gives place to discussions regarding epistemology and no provided consensus. While some naturalists argue that epistemology should be abandoned to science, others argue that naturalistic epistemology should be guided by scientific methods and findings (Clarke, 2009). Philosophical naturalism attributes verification based on scientific methods to understand natural phenomenon (Raleigh, 2022).

Empiricist verification, on the other hand, suggests that sensory experience is essential for knowledge (Westphal, 2017). In this respect empiricist verification is foundationalist and demands that knowledge should be relied on as a secure (i.e., permanent, indisputable) foundation (Schwandt, 2015). Empiricist approaches also suggested that "for any non-logical proposition which can be known to be true, there is some possible sensory experience that would (partly) confirm the proposition" (Westphal, 2017: 9).

Another approach is logical empiricism (or *logical positivism*) (Neuber, 2014). According to logical empiricism, verification can be achieved through analytic methods such as analytic studies (i.e., logic and math) or empirical studies. As well as being an ontological subject, verification is also related to meaning, and in this respect verification is also a semantic issue according to logical empiricism (Cummings, 1998; Ney, 2012). Logical empiricists attribute importance to mixed methods for a robust form of verification. The main reason for this stand lies in the epistemological view of logical empiricism. Logical empiricism aims to provide access to factual truths by allowing methodological flexibility in social and behavioral research (Schwandt, 2015). Logical empiricism seems to have been transformed into *statistical empiricism* in our present scientific world. According to this approach, statistical findings are the basis of verification (Papageorgiou and Lekkas, 2020).

Since scientific pragmatism accepts reality based on the subject's experience, i.e., the act of verification has meaning in terms of each actor's subjective reality. Verification in scientific pragmatism is condition-dependent, and evidence may not be sufficient for verification. This is the main difference between scientific pragmatism and other approaches primarily.

Philosophical naturalism assigns verification to scientific methods. On the other hand, while empiricism provides verification based on sensory experience, the approach to verification of logical empiricism considers the meaning of propositions as well as flexible methods. As can be seen, there is no consensus on how reality is to be accessed or even how crucial verification is in the scientific world.

Epistemology mostly focuses on propositional knowledge, namely knowledge of facts (also known as factual knowledge), the knowledge that something is so, and the knowledge that some proposition is true (Audi, 2010). For example, "I know that bears are mammals." Is a factual knowledge (Rescher, 2003). Epistemological questions arise under some intuitional conditions. Intuition can be *a priori* for conceptualization (Cruz, 2006). A priori refers to a kind of justification that typically does not require evidence from sensory experiences (Moser, 2005). But in a conceptual sense, a priori toward epistemological questions can derive from the senses, reasoning, and beliefs (Cruz, 2006).

Epistemological verification can be *strong* or *weak. Strong epistemic verification* indicates the acceptance of a belief as definitively true. *Weak epistemic verification* denotes that one has granted some credit to a belief, seeing it as a plausible prospect (Rescher, 2003).

Epistemology is *normative* and focuses on how people should reason and what they should believe (Bishop and Trout, 2016). Two types of theories dominate epistemology. *Belief theories* aim to explain a normative category that applies to individual beliefs. *Theories of knowledge* and *justification* are belief theories. *Reasoning theories* aim to explain a normative category that applies primarily to reasoning, whereby that reasoning typically leads to beliefs. Theories of rationality are reasoning theories (Bishop and Trout, 2016).

Epistemological problems that can be encountered in scientific research can be grouped under three headings (Hon, 1989; Rescher, 2003):

1. **Errors of reasoning**

→ *Poor or wrong reasoning* (Rescher, 2003): Prevalently, scientific inquiry processes exhibit complex courses of reasoning. Correct and effective reasoning is very important to accomplish scientific inquiry. The reasoning process is related to consideration of correct theories, choosing relevant variables for testing/measurement, modelling, and drawing inferences from results.

2. **Experimental/observational/measurement errors**

→ *Experimental errors* (Hon, 1989): This problem may start with choosing the wrong background and/or instrumental theory. A false general and/or instrumental theory leads to epistemological problems in any scientific inquiry. Experimental errors may also reflect assumptions concerning the actual set-up and its working standards, observational reports, and theoretical conclusions.

3. **Wrong inferences**

→ *Poor presentation of the evidence* (Rescher, 2003): This epistemological problem is related to weak epistemic justification.

Briefly, in the context of epistemology, discussions in the philosophy of science are centered on *verification, logical foundations, meaning,* and *truth.*

SCIENTIFIC ETHICS

Although the words *morality* and *ethics* are used interchangeably, there is an important difference between the two. In general, **morality** consists of the principles or rules that define what is *right* and *wrong*. For example, principles such as "*it is wrong to lie*" and "*it is good to help others*" point to morality. **Ethics** is the process of examining moral standards and how these standards should be interpreted and applied in various situations (Weathington, Cunningham, and Pittenger, 2012).

Scientific research should aim to minimize risks and maximize benefits. Thus, scientific activities are not free of ethical concerns. Scientific researchers also have ethical responsibilities throughout the entire research process. Ethical responsibility, in terms of the research process, is broad, from the process of collecting data to analyzing and interpreting them. Ethical responsibility requires data to be gathered without posing any risk to the safety, well-being, and health of people, living beings, and the environment, as well as not manipulating the data during analysis, and considering the ethics of science in the use of knowledge (e.g., industrial applications). In short, scientific ethics covers every stage, from collecting and processing data to the use of knowledge (Abrams and McMillan, 2015).

Researchers working in fields such as medicine and psychology may need to show more ethical sensitivity than in many other fields since they can directly affect the lives and health of people and living beings. For example, studies involving children should not risk their health and development. Further, children are more vulnerable, have fewer legal rights, and often do not understand the concept of *informed consent.*[13] When conducting research on children, researchers must obtain permission from parents or legal guardians for children to participate (Abrams and McMillan, 2015; Fraenkel, Wallen, and Hyun, 2011; Weathington, Cunningham, and Pittenger, 2012).

Apart from these scenarios, we should keep in mind that every field of science may require specific ethical codes. In professional practice, professionals are subject to professional ethical codes, which are sets of rules they must follow when performing their jobs. In this context, it is beneficial to follow ethical standards that developed in light of philosophy (Abrams and McMillan, 2015; Fraenkel, Wallen, and Hyun, 2011; Weathington, Cunningham, and Pittenger, 2012). Moral standards that can be a guiding light in terms of scientific research include the following (Weathington, Cunningham, and Pittenger, 2012):

- ✓ The well-being of the people affected by our behavior/actions is important.
- ✓ Moral principles transcend personal standards such as self-interest.
- ✓ Moral principles are universal and persistent.
- ✓ Moral principles are neutral.

These were published in the Belmont Report (1979), an ethical guideline for human protection, especially for medical and behavioral research (Rosenstein, 2019). However, looking further in the past, we find the Nuremberg Research Ethics Principles (1947)[14]:

- ✓ Voluntary consent is essential.[15]
- ✓ Study outcomes should be for the good of society.
- ✓ Scientific validity: Research should be done carefully and be well-designed by trained scientists.
- ✓ The research should be designed with an understanding that maximizes safety and minimizes risk.

---

[13] Informed consent is prepared in written form. Consent documents should contain the following characteristics: (1) Potential participants should be provided with all necessary information so they can decide whether to take part; (2) potential participants must understand the information that describes the research and what participation will involve; and (3) the voluntary nature of participation must be made explicit (McMillan, 2015: 34).

[14] Cunningham, Weathington, and Pittenger, 2013; https://encyclopedia.ushmm.org/content/en/article/the-nuremberg-code

[15] In some cases that require passive observation of people's behavior in the context of natural flow, data may be collected without asking permission.

- ✓ The research should be canceled if there are conditions that may cause physical or mental harm to participants.
- ✓ This should also be applied when the cancellation request comes from the participant.
- ✓ The researcher must protect the participant's privacy.
- ✓ Researchers should take special care to protect the rights of children and those who are developmentally delayed or who cannot move by themselves (vulnerable populations), including those with mental disabilities.
- ✓ While choosing participants for the research, fairness and consistency should be ensured.
- ✓ The researcher should constantly monitor his/her study to guarantee the safety of the participants.

American Psychological Association (APA, 2020: 254) defines plagiarism "*the practice of presenting the ideas or words of others as one's own without appropriate acknowledgment of the source*". **Plagiarizing** involves copying another person's ideas, words, or work and pretending that they are your own. In other words, plagiarism is the misappropriation of someone's work without referring to the original source. Plagiarism is an ethical problem for both students and professional researchers. Unintentional plagiarism is also plagiarism. Even failing to cite a small number of sources is plagiarism. Plagiarism violates scientific norms and can be regarded as theft of another's work. Making appropriate citations and referring to the work(s) one has used in research is crucial in terms of scientific ethics (Abrams and McMillan, 2015; Masic, 2020).

Various factors cause plagiarism, such as a poor academic background, a lack of writing skills, and ignorance of respecting intellectual property[16] (Masic, 2020). Plagiarism can occur in several forms; one type is *self-plagiarism*, which means taking paragraphs from one's own work. Researchers should cite references to their own work(s) to avoid self-plagiarism (Abrams and McMillan, 2015).

There are some tools to detect plagiarism. For example, **iThenticate®** is a program used to scan published works (e.g., articles, books, dissertations, etc.) and to detect the matching parts of an article. IThenticate® is only used for articles. **Turnitin®** is another program used to scan all academic works (articles, books, dissertations, and student assignments) and to detect the matching parts of course assignments and dissertations (Masic, 2020).

---

[16] Intellectual property is protected by copyright law. For this reason, plagiarism also violates the law.

Deception is an important ethical problem. **Deception by omission** occurs when facts about the research are hidden from the participant. This problem can be encountered in almost every kind of research since it is not always possible to share the full content and purpose of a study with the participants in detail. In the case of omission, the contribution that the participant could provide may be invalid, or the participant could experience mental ambiguity regarding the content or purpose of the research. Deception (and deceit) by **deliberately lying** or **misleading the participant** about an important component of the study is another ethical problem (Cunningham, Weathington, and Pittenger, 2013; McMillan, 2015).

Another ethical concept related to scientific research is **anonymity (or confidentiality)**. Participants are expected to share their views, perceptions, or thoughts accurately and honestly. Protecting a participant's confidentiality and ensuring the privacy of personal information are essential for pursuing scientific research. Participants' privacy must be protected and collected data must be anonymous (McMillan, 2015).

Finally, scientists are expected not to comment on scientific subjects in which they have any kind of personal interest. For example, it is against scientific ethics for a scientist to express an opinion on the public health effects of the drugs of a pharmaceutical company in which he is a shareholder or that is a sponsor for her/his scientific studies.

CONCLUDING REMARKS

I consider that it is necessary to recognize the philosophy of science, logic, epistemology, and ethics to a certain extent before learning scientific research methods. Because the mind needs the guidance of epistemological standards, logic, and philosophy. Additionally, researchers should not conduct science without ethical standards.

SUMMARY OF CHAPTER 2

The *philosophy of science* is grounded in the philosophical explanation of science and guides scientific consistency, internal integrity (e.g., in terms of modeling), and the production of reliable outcomes. Classical and modern approaches to the philosophy of science vary but some principles and assumptions are universal.

Human mind works on a *logical basis*. Universal standards of logic are independent from the human mind and we have to learn how to use those standards. Logic provides us with consistency in scientific studies.

Deduction and *induction* are both ways of inferencing on a logical basis. Logical-based cognitive skills, which we also use in our daily lives, become even more essential of an issue to avoid making wrong inferences while conducting deduction or induction operations in scientific studies.

Epistemology plays a philosophical guiding role on how researchers can reach truth/reality, especially with its verification approaches and standards. Epistemology is the science of knowledge and one of the subjects of philosophy. A researcher should assure the knowledge produced is reliable and valid. For this reason, science practitioners should recognize the types of knowledge and features of reliable and valid knowledge. In a technical sense, a researcher should assure the data she/he uses are clear and collected according to scientific necessities. Besides, epistemological verification can be strong or weak. Because of this, some assertive conclusions may need to be verified by the repetition of surveys and analysis.

Scientific ethics is another important topic that should have been understood well. Researchers should follow the ethical standards of pursuing science. Ethical standards can also be extended to professional areas when it comes to application. Practitioners should recognize and pay attention to the ethical standards of their professional fields.

Chapter 3

ONTOLOGICAL ASPECT OF RESEARCH SUBJECTS

Before starting to read the chapter

We live in a world consisting of *entities* that have *properties.* Those entities also have *relations* with one another. Science examines those entities, properties, and relations. Because of this, it is crucial to understand this concept of entities. Entities are the subject of ontology. This chapter presents a brief statement of *ontology* and *existence* and attempts to answer questions such as -*What is existence?*-. This chapter also gives a place for the notions of *phenomenology* and *reality.*

It is important for readers to comprehend *our world of phenomena* composed of facts, objects, and events ontologically. Thus, readers will be able to distinguish those phenomena with their *realities, properties, and relatedness,* which provide the basis for variables.

Ontology of scientific subjects necessitates comprehending the *properties* of studied entities by researchers. It is also important for researchers to gain an understanding of *randomness* to understand variables, which are discussed in Chapter 5 in detail.

Another concept regarding the world of entities is *relatedness,* which is closely relevant to the term *relation.* Relations have many variants, such as *association, interdependence,* or *causality.* All those notions are discussed under the sub-section of *relatedness* in this chapter.

ONTOLOGY OF SCIENTIFIC RESEARCH SUBJECTS

We live in a real-world that is independent of—and external to—the human mind, and, therefore, its experience or cognition. This independent world consists of *entities, properties* belonging to those entities, and *relations* between those entities (Bryant, 1997; McMillan, 2015; Outhwaite and Turner, 2007). However, only observable and measurable facts, objects, and events become subjects of science. Facts, objects, or events that are the subjects of scientific research have many *characteristics (features)* and can be the subjects of different research models and designs according to their own *characteristics* or their *relations* with their surroundings.

Ontology is the science of being and concerns the categorization of what exists (Rosenkrantz, 2012). Scientific research always seeks *reality*, which exists ontologically. Reality is about facts, objects, and events, because only entities that can be observed and measured can become subjects of science. Many different perspectives can be developed according to those facts, objects, and events. These varying perspectives enable one to study diverse subjects of research. In other words, we are referring to the time horizons (past, present, and future) of facts, objects, and events; situations in space (position); actions (such as active or passive, static or dynamic); structural states (forms that are similar or different, size [large or small], conditions [flexible or rigid, sparse or intense]); environmental conditions (surrounding elements); components (sub-elements, sub-dimensions, or functions); internal processes; and contexts (relationships, effects, communication, the exchange of materials and interactions). The facts, objects, or events that are addressed can be examined regarding one or more of these features.

Let us try to illustrate the information described above with one example from the natural sciences. We will take the Sun as an object.

> The Sun, as an object (which is also an ontological being), is located in the inner neighborhood of a spiral arm within the Milky Way (the location state in space/macro focus), is about 4.6 billion years old and has completed half of its life span (a time horizon), is medium-sized (structural state), and is an active (action state) entity. It is a star in plasma form (structural state) and has a spherical shape (structural state). Eight planets are associated with the Sun (the microfocus environment). These planets (peripheral elements of the Sun) follow an elliptical orbital[1] movement around the Sun. The Sun, along with neighboring star-groups within the Milky Way, moves in an orbit around the galaxy. It produces energy (output) by converting hydrogen atoms (components-input), which are part of its content, into helium (components-output) through fusion (an internal process) under high gravity. The Sun, which consists of a series of layers (structural sub-dimensions) starting from the nucleus, has dark spots on its surface, where the force of magnetism is concentrated. The Sun also applies enough gravitation (effect) to keep the planets in its orbit. The planets, on the other hand, have a certain resistance to this gravitation; they avoid being drawn into the Sun via sideways motion and follow trajectories determined by the balance (through an interaction) between their force of motion and the gravitation of the Sun.

---

[1] Venus is an exception; Venus follows an almost circular orbit.

In this example from astronomy, we tried to explain some factual information about the sun in the context of its *state in macro focus* and *microfocus, structure, components, actions,* and *internal processes,* based on its location in the Milky Way Galaxy. The purpose of giving this example is to show that factual realities can be considered in many contexts.

Thus, scientific research is an ontological study within the framework specified in the example above, and hypotheses developed in accordance with accepted theories and logical bases can be tested through research.

If we want to expand our examples, earthquakes and mortality are also facts. However, if we are talking about a specific earthquake, the earthquake here is an event. Similarly, if we are talking about the mortality process of a specific living being, mortality is an event here. Phenomenologically, earthquakes and mortality point to a general phenomenon. Science can examine these realities as facts or as individual events (Smith, 2018). The content of this chapter addresses the *entity, properties,* and *relatedness* of the research subject.

The Entity Dimension

Our world comprises material *objects*[2] (e.g., trees, buildings, soil, stones, flowers, the sky, etc.), which are tangible, i.e., we can experience them through our senses. However, the world in which we live is not solely comprised of material objects. There are also immaterial or abstract objects, such as properties, states, and relations accompanying material objects.[3] We also perceive, comprehend, and experience *events*[4] (also known as *occurrence,* e.g., running, learning, a working production process, ongoing wars, etc.) and *facts*

---

[2] Some authors (e.g., Easton 2010) use the concept (or term) of the object with the same meaning as the concept of the entity containing events (e.g., Rodrigues and Abel, 2019). I preferred to narrow the concept of the object to make a distinction in this book.

[3] The concept of the object involves some related abstract and immaterial terms such as *quality/property/feature* (e.g., material form, propositions, ideas, etc.), *situation/state* (e.g., sets, spatiotemporal positions), *relations* (Rosenkrantz, 2012; Galich and Shnyakina, 2015; Rodrigues and Abel, 2019). As can be seen, it is not easy to come to a consensus on concepts that seem easy at first glance in philosophy. For this reason, the subject of semantics (see Chapter 4) is included in this book. Semantics is an important field of study both for linguistics and philosophy. Thinkers and authors consult linguistics to solve semantic problems.

[4] *Events* ontologically are associated with time. More clearly, events happen in time by accumulating different temporal parts (Rodrigues and Abel, 2019). Some authors distinguish between the concepts of event and process. I preferred using the concept of the *event* to refer to any kind of happenings such as *process,* or *activity,* and besides, using it for the same meaning as *occurrence.*

(e.g., birth, death, gravity, etc.). Every fact, object, and event[5] surrounding us is an *entity* and the subject of ontology and existence (Craig, 2005; Mellor, 2005; Rodrigues and Abel, 2019, Rosenkrantz, 2012). In other words, ontology examines entities. Those entities exhibit (or have) several characteristics (features or qualities). Entities (or truths) can be *particular* and *universal, concrete* and *abstract,* or *contingent* and *necessary.* There are many other classifications of entities presented in ontological studies, and there is no agreement on an absolute classification of entities (Cowling, 2017; Easton, 2010; MacBride, 2005; Rosenkrantz, 2012).

Ontology and Existence

The cosmic universe contains everything we know. This hosting (or including, enclosing) also refers to a scope. *What does the universe consist of?* The first possible conceivable answers to this question are galaxies, stars, planets, gas clouds, and asteroids.

When philosophers discuss existence, they generally mean *physical existence* in the natural world. This view of existence can be described as follows (Rescher, 2003):

(1) The things we experience with our internal and external senses exist.

(2) The things whose existence we need to postulate in order to realize an adequate causal explanation of things that exist also exist. This causal explanation also indicates the linkages of cause and effect between entities.

The objects mentioned above are the classes of celestial objects due to a systematic distinction that we make. What would we say if we wanted to go beyond the boundaries of our classification and look at the universe a little more ontologically? Let us scrutinize the concept of ontology now.

Ontology is the branch of science that examines existence, the existences that exist, and what sorts/types of things exist. The question of ontology is "*What is it?*" (Smith, 2018). This aspect is also related to real things (Glesne, 2011).

Ontology was called by Aristotle as *the first philosophy.* Because, according to Aristotle ontology was "the philosophical study of being in general or of what applies neutrally to everything that is real" (Britannica, n.d.). In this respect, ontology is related to reality, and ontological thinking is the effort to "describe

---

[5] In this book, the terms *fact, object,* and *event* are used to make a distinction between entities. Here, the three terms, together, reflect the concept of the entity. Of course, this distinction can be found restrictive by readers. But my aim is to distinguish and simplify the concepts as much as possible.

and explain reality in the only correct and binding (for everyone) way" (Matuszek, 2015: 204). Ontology appeals to particular distinctions for reality, such as "system/environment, operation/observation, and reality/construction" (Matuszek, 2015: 208). Philosophical questions about existence include those concerning concrete and abstract things. We can think of the following as a mental exercise. We can follow concrete examples, such as "*Are there other universes?*" and "*Is there substance?,*" or we can follow abstract examples, such as "*Is there a mind or a soul?*" You might also ask, "*Is there only substance?*" and "*Are there things that exist regardless of space and time?*" Hence, it is possible to generate questions about existence at very different levels and in different contexts[6] (Craig, 2005).

How can we examine the universe after these explanations? Let us try to answer the question "*What does the universe consist of?*" using our knowledge of high school physics: The universe consists of matter and energy. It seems plausible!

A concept closely related to ontology is **existence**, which is a widely discussed topic in philosophy. It is usually handled in two contexts. In the first, we seek to answer the question "*What is existence?*" In the other context, we examine the question, "*What is there?*" These discussions also explore existence regarding **properties/characteristics** and involve concrete or abstract existences (Mackie, 2005).

Phenomenology

The lifeworld is a world as it is experienced. This notion is phenomenologically true because phenomenology indicates that a phenomenon is an entity as it is experienced (or as it is understood) by humans. In other words, **phenomenology** is the investigation of phenomena. A phenomenon may be an object, a natural or social fact, or an event. We come across uses in the forms of natural or cultural phenomena. For example, *globalization* is a social phenomenon (Dahlberg, 1995; Oxford Dictionary, n.d.).

Phenomenology can also be defined as the study of structures of experience and consciousness. In this respect, phenomenology is directly tied to human experience and consciousness since objects and events are experienced by people primarily through the senses and interpreted cognitively. Of course, human experience pathways cannot be reduced to the senses, such as touching, hearing, and seeing, as human experience is much richer than that. Phenomenology examines the structure of a wide variety of experiences,

---

[6] The main philosophical currents that address ontology are *existentialism, materialism, idealism, realism,* and *nominalism* (Craig, 2005).

encompassing perceptions, thoughts, memories, imagination, emotions, desire and will, physical awareness, concrete actions, and social (including linguistic) activities (Smith, 2018).

Phenomenology and ontology are closely related scientific areas, but are not the same. As explained under the previous heading, ontology investigates existence. Phenomenology examines the experience of what exists, not what does not exist. Therefore, the two concepts have close—but different—meanings. Phenomenology is a field of philosophy focused on how a fact, object, or event is experienced by humans, rather than the reality about that fact, object, or event (Oxford Dictionary, n.d.; Smith, 2018).

If we continue from the example of the universe, our universe covers many objects and events. The Big Bang is an ontological event; our resulting universe is an ontological being. A star itself is a being; a star's death is an event. Let us remember that a particular mortality process is an event and phenomenon, while mortality is a fact, like a general truth. Similarly, gravity is a natural fact like mortality. Their being points to an ontological state. All these facts and events are perceptible to the senses, and at this point, they become the subjects of phenomenology. Also, if we are talking about objects, we are talking about the forms that characterize them.

Reality

Reality is the truth about a fact, event, or object. There are many views of reality. Some philosophers consider reality to be a process, some consider it a static state, and some see it as *flexible,* while others relate it to human perceptions or language. Some philosophers assert that we construct reality in the mind, and some philosophers claim that reality is independent of all cognition and language; in other words, reality is fixed and independent of any human understanding (Bryant, 1997). According to some philosophers, reality can be modeled through set-theoretical models in terms of *entities, properties,* and *relations* (Bryant, 1997).

Reality has been and is still discussed by many philosophers in the past and today. The ontological and operational meaning of the concept of reality may differ for many approaches. For example, from a constructivist perspective, there "is a distinction between reality independent of observation and reality constructed in observation" (Matuszek, 2015: 204). Regarding this distinction, Matuszek (2015: 205) stresses that "the point here is not the division into ontological reality and constructed reality, but that the distinction itself is ontological, regardless of whether a reality independent of observation does exist or does not." According to the constructivist approach it is impossible to say anything about reality independent of observation (Matuszek, 2015: 208).

Empiricism (Schwandt, 2001) and naturalism (Rouse, 2007) also assert that reality is independent from the observer. Those approaches distinguish ontological reality and constructed reality in mind. According to those approaches, we construct reality based on our senses.

In ancient Greece, reality was determined through individual observations. A theory was developed after observations (Stephens, Barton, and Haslett, 2009). However, there is a relationship between religious, cultural, political, economic values, and scientific judgment. In other words, the attitudes and prejudices (stereotypes) we maintain in the social context can affect the scientific research process. Therefore, self-awareness is crucial for people engaged in science. It is important for scientists to recognize themselves and to understand the prejudices or distortions caused by their religious values, economics, political views, and cultural features. Science requires us to be objective (See Chapter 4). Objectivity is possible by pursuing truth, independent of the researcher's own beliefs and worldview (Beach, 1996).

Epistemologists have pointed out the famous two-table discussion of the British physicist-astronomer Sir Arthur Eddington, who, regarding views of reality, contrasted the solid table of everyday experience with the physicists' table of a manifold of electromagnetic oscillations in primarily empty space. The latter is the real table as it exists in nature, while the first table is merely an appearance, a delusion, "*a mirage, as it were*" existing in people's minds. Our ordinary life view of the world is a matter of mental (rather than optical) illusion (Rescher, 2003).

At this point, it is necessary to talk about realism. The basic idea of **realism** is to examine what kinds of things exist (as existentialism does), what things do not exist, or what they differ from. Many people find it natural to be realistic. For example, how many planets are there in our solar system is independent of how many planets we believe there are. Therefore, a realistic response will be independent of the subject (Craig, 2005).

> **"Extraordinary claims require extraordinary evidence."**
>
> (Carl Sagan 1979)

The Properties (Features/Characteristics) Dimension

A **property** (sometimes called a feature or characteristic) refers to a variable of a subject examined in research. Properties ontologically arise from structures, processes, and components. In this respect, they also determine the boundaries of an ontological being. This is a very important concept to understand comprehensively since all scientific research handles the subject studied in terms of its various properties. I explain the variable concept, which is related to properties, in Chapter 5 in detail. I discuss the concept of randomness under the following title.

Randomness

The fundamental view underlying **randomness** is the absence of order or patterns. Scientists and philosophers apply five approaches to randomness (Dembski, 2005):

(1) Randomness is the output of a process based on chance. Thus, an event is random if it is the output of an activity based on chance.

(2) Statisticians view randomness a way to obtain a highly representative sample according to a specified probability distribution. In other words, statisticians prefer random samples in the sense of the process being based on chance. As a result, a sequence of chance-based sampling processes constitutes a *random sample.*

(3) There is a given mixture of particles in a fluid in a chemical experiment. Eventually, the particles become thoroughly mixed throughout the fluid, reaching an equilibrium state. Hence, randomness is identified as the equilibrium state reached via mixing.

(4) Randomness is a measure of computational complexity in computer science.

(5) Finally, randomness implies pattern-breaking. If an object is random in a specified collection of patterns, the object breaks a pattern. Here, it is critical for the object not to fit with even one pattern in the collection. Otherwise, the object cannot be identified as breaking a pattern.

Randomness is closely linked to the working of nature. Because of this feature, there is an association between randomness and *complexity.* This association also indicates that randomness is close to *chaos* (Sawyer, 2007).

As seen in the second approach presented above, randomness is not only related to the work of nature; it is also tied to *methodology.* According to the *law of large numbers,* any large number of items chosen randomly from a population will have, on average, the same characteristics as the population (Weathington, Cunningham, and Pittenger, 2012). It is crucial to be aware that random sampling can be suitable for a quantitative research study. On the other hand, a qualitative research study mostly necessitates selecting a sample purposefully (Gay, Mills, and Airasian, 2011).

The indeterminacy of modern physics seems to be due to pure randomness. Some scientists and philosophers assume that randomness directed by chance can be calculated via math and statistical methods. As is known, we use those methods in many scientific and professional areas. However, the question is: To what extent? According to the perspective of modern physics, and particularly of quantum mechanics, randomness seems to follow intricate paths. The readers of this book can trace the concept of randomness in light of recent developments in physics, particularly those related to quantum mechanics.

The Relatedness

Before I start, I should point out that the title Relatedness was my choice, and I collected related terms under this title because there is no consensus on those concepts in terms of meaning or the typology of relatedness.

The Oxford Dictionary (n.d.) defines **relatedness** as "the fact of being connected with something/somebody in some way." Scientifically, relatedness refers to **context**, which entails all elements being linked to each other in a fact, situation, event (process), process, or object. For example, let us imagine a production activity that itself consists of many operations (or processes). Each of these processes (stages) constitutes the content of this activity. These processes are systematically linked (in a logical and scientific-systematic order) in our production activity, leading to the final product. Therefore, each process is the output of the previous process and the input of the next one. The systematic connection among processes is structured in a particular *context.* The explanation of *why* and *how* two processes are tied to each other is the *context* of those two processes.

Contexts can appear in different forms, such as *interest, relatedness, situation,* and *sequence.* As a result, concrete and measurable words such as *definition, explanation, comparison, ordering,* and *classification* are included in scientific research in terms of context. Immeasurable, abstract, and indefinite verbs are useless for scientific research. In addition to quantitative research results, qualitative research results are also expected to produce comparable or classifiable outcomes. I discuss the matter of context in detail in Chapter 4.

We indicate the concept of relatedness in the sense of scientific research, and particularly to the connections between facts, objects, events, and variables studied. It is important to distinguish between context and relatedness. Relatedness is about the connection type of subjects studied. There are many types of relatedness between facts, objects, and events, such as an *association, relationship, relevance, effect, interaction,* and *interdependence,* which I discuss under the following heading. Sometimes it can seem hard to distinguish the terms found in the title below. Some terms seem to have close meanings to one another or are almost synonymous. Here, I will present a challenge to make them much clearer.

Associations, Relationships, Relevance, Effects, Interactions, and Interdependence

In our daily lives, we make connections between events and observe relationships (or associations) between various elements. We do this by processing the data from our senses in the mind. These connections and relationships appear to us as facts about the world we live in (Fraenkel, Wallen, and Hyun, 2011). For example, let us consider a manufacturing activity. A production activity consists of many stages (or processes). Each of these processes (or stages) forms the content of the activity. These processes are systematically linked (based on a logical, scientific order and method) to the production activity, leading to the final product. Therefore, each process is the output of the previous one and the input of the next process. The systematic connection among processes embodies the context; that is, the explanation of why and how the two processes are related is the context of these two processes. From a scientific angle, scientific research can reveal the variables that are the subject of the research in various contexts.

Just as in our daily lives, researchers also use words such as *relevance, effect, interaction,* and *relationship* more carefully, remaining loyal to their exact meanings in a scientific study. Even if some words sound very close to each other and seem interchangeable, they might not be. Because of this, researchers try to choose the terms that fit the best. Let us examine at how the Cambridge Dictionary (n.d.) defines the terms we mentioned in the title:

- ⇨ **Association** is a connection or relationship between two things or people
- ⇨ **Relation** is a connection between two or more things.
- ⇨ **Relevance** is the degree to which something is related.
- ⇨ **Effect** is a change, reaction, or result caused by something.
- ⇨ **Interaction** is the way that two or more things combine with and affect each other.
- ⇨ **Interdependence** is the fact of entities depending on each other.

The relevance of facts is generally related to a coherent system of (natural or social) laws (Bohm, 1961). Relations between two or more variables also indicate functions of variables. Those relations (interactions) are based on causality in social and natural systems (Mulaik, 1986). Associations do not always refer to functional relations. Associating variables may have a common cause reflecting premises that need to be tested (Mulaik, 1986). Interdependence concerns the dependency of elements such as social groups and economic systems. Interactions are common in natural and social life. For example, humans experience interactions with their environment (Weathington, Cunningham, and Pittenger, 2012). As can be seen, some concepts seem to be intertwined in terms of meaning, such as relation and association.

Relationships can be *hierarchical* or *dynamic* in terms of their features (Marvin, 2011). For example, relationships between management units of an organization or thermodynamic systems can be hierarchical. On the other hand, market relations in a national economy exhibit dynamic features. Similarly, the relationship between the physical and psychological states of people exhibits dynamic features.

Relationships can also be *linear* or *non-linear.* A *linear relationship* refers to the increase (or decrease) in one variable and the increase (or decrease) in another variable. For example, converting the measure of length from mile to kilometer exhibits a linear relationship. A *non-linear relationship* (a curvilinear curve) shows that the distribution of the relationship between two variables is not straight-lined; that is, it follows a curved line. For example, the relationship between temperature and aggression will follow a curvilinear line. Suppose that temperature increases anger after exceeding 21 °C, and that the level of anger decreases after 32 °C (Fraenkel, Wallen, and Hyun, 2011; Mitchell and Jolley, 2009). Non-linear relations especially prevail within human systems and behaviors due to their complex dynamic nature (Jura et al., 2006).

Interactions can also be seen where multiple independent variables have an effect on the variable that is dependent on the interaction. In other words, an interaction can occur between a dependent and an independent variable, or between two or more independent variables. The interaction shows the degree of differentiation of one independent variable from the others between the levels of the independent variables (Rosenstein, 2019). An interaction means that the effect of one independent variable is not consistent across all levels of the other independent variables. An interaction can also indicate a unique effect that neither independent variable can explain by itself. An interaction between independent variables is determined by *factor analysis* (Weathington, Cunningham, and Pittenger, 2012).

Causality (Cause and Effect)

In the scientific world, the cause-effect relationships between variables, as well as components, processes, and conclusions about facts, objects, and events, are widely examined. Causality is a particular type of relatedness between two or more variables and is widely studied in many scientific areas such as medicine, psychology, biology, sociology, economics, etc.

Carnap (1928) stresses causality as a functional dependency between variables (Mahoney and Acosta, 2022). Causality is not as simple a type of relationship as it seems. Causality needs scientific evidence and correlation is a useful piece of evidence to establish causal relations. But correlation alone does not always indicate causality. In other words, correlation does not guarantee causality (Russo, 2017). Two variables can exhibit high correlation but it does not mean they are related. The correlation between them can be a coincidence. Because of this, the logical basis for causality, supported by correlation, becomes very important. For example, a correlation can be found between drowning at sea and ice cream consumption. This does not mean that ice cream consumption increases the incidence of drowning at sea. There is only a seasonal coincidence. As can be seen from this example, causality is a logical argument.

Another issue related to causality is the possibility of the availability of multiple causes that can play a role in a particular effect. In other words, an effect can arise as the production of one or more causes. But those causes have different roles in that particular effect. Causes can be *necessary* and/or *sufficient* (Russo, 2017). For example, smoking is only one component of the complete sufficient cause of lung cancer. There must exist other causes to arise effect. Let us examine another example to fully understand the roles of necessary and sufficient causes. Suppose Alice's house burned down, and the fire was caused by an electrical short circuit. It can be seen that an electrical short circuit is not sufficient to cause the fire all by itself. There must necessarily be another cause, such as oxygen. Of course, investigators report that the cause of the fire was the electrical short circuit, not oxygen (Illari and Russo, 2014).

In contemporary research, it is common to determine cause-effect relationships based on the criteria set by John Stuart Mill (1806–1873). Mill argued that three indicators should be looked at to find a cause-effect relationship (Weathington, Cunningham, and Pittenger, 2012):

- ✓ In terms of time order, the cause must have appeared before the effect.
- ✓ There should be a correlation between the two events/variables.
- ✓ All other possible explanations for the relationship between the two events/variables should be either ignored or explained.

The third criterion is especially difficult to meet, particularly when searching for a cause-effect relationship, as it is often difficult to consider all possible explanations that initiate an effect (Weathington, Cunningham, and Pittenger, 2012).

In the contemporary approach, there are three criteria for causality: (1) there must be a strong correlation between cause and effect variables; (2) "the proposed cause must precede the effect in time"; (3) the cause variable must be present during the effect (Grove and Gray, 2018: 501).

Woodward (2010) discusses causality in terms of interventionism. Woodward points out causal relationships have the potential for an exploitable relationship for purposes of manipulation and control. A dependent variable of a causal relationship can be changed by the manipulation of the independent variable. Those discussions also bring us to the *free will* problem of subjects/actors because; manipulation of subjects is a problem for causal determinism (see Chapter 1).

The free will notion as the causal source is also related to *moral responsibility*. One view suggests that a manipulated subject does not have responsibility for her/his actions because subjects act in accordance with the manipulator's intentions. Due to not being the causal sources of their actions, they are not responsible for their actions (Deery and Nahmias, 2017).

Another view of causation asserts that the possibility of *preventions* and *exceptions* are natural parts of causation. This view considers causation as a disposition toward an effect that is evitable (Mumford and Anjum, 2010).

Causal relationships are initially conceptualized in the mind. We can try to express the causality as an equation, diagram, or equivalent verbal assertion (Hayduk and Pazderka-Robinson, 2007). For example, to understand the relationship between cost and demand variables, we can express them in a mathematical equation. If causality is to be presented visually via a diagram, certain steps can be followed (Creswell and Creswell, 2017):

- ✓ The position of the dependent variable should be on the right in the diagram, while the position of the independent variable should be on the left in the diagram.
- ✓ A one-way arrow should be used to indicate the effect of each determining variable in relation to each variable that is dependent on it.
- ✓ *independent variable A → dependent variable B*
- ✓ The strength of the relationship among the variables can be shown by inserting positive or negative signs on the paths.

Causes can be related to events (consequences) through logical inferences or probability formulations. Causal variables might not be directly observable; thus, they may not have been included in the research model. In such cases, predicted associations between observed variables are not sufficient to explain the causal relation. This kind of lack can be detected via evidence (Mulaik, 1986).

If we heat water, it boils after reaching a given temperature. This happens as a function of raising the temperature of the water. It is an example of a direct effect. However, we might experience indirect effects in life. For example, consuming a high level of sugar can cause a person to become diabetic. Having diabetes can lead to pancreatic cancer. Hence, there is an indirect relationship (effect) between sugar consumption and pancreatic cancer. Indirect effects can also be observed in social events. Many social (as well as economic and political) events have indirect effects on new events or conclusions. For example, stagnation can cause unemployment, and unemployment can change a country's political conditions, such that stagnation can indirectly impact a country's political conditions.

Causal relations might be tricky in some cases especially when we talk about social issues. Often causal relations may seem easy despite they are not, and sometimes vice versa. For this reason, researchers should pay attention and follow logical clues when establishing causal relationships.

CONCLUDING REMARKS

In my personal observations in my professional life, I noticed that young researchers had difficulty in grasping the ontological features of the scientific subjects being studied. This situation was due to not knowing what the ontological features that are the basis of the concept of variable were. This chapter is important as it prepares the readers for the concept of variables in Chapter 5.

SUMMARY OF CHAPTER 3

Science is the sum of our efforts attempting to understand ourselves, our societies, living populations, the world, and the universe. Research studies frequently examine properties that change over time, and we call them variables. In other words, all *entities*, whether living creature or objects, has variable *properties*, and *relates* to their environment.

Variability is related to time - covering the past, present, and future. If we know how a property changed from past to present, we can attempt to make a prediction for the future. Geologists strive to predict the time range or the intensity of the next earthquake by looking at the data of past earthquakes for a specific location. Marketing experts seek to predict market growth for the following years through the past data of the market. Astronomers seek to predict the next arrival of a comet. All these examples point out that those entities can be examined based on all their variable properties and relationships. This is where the models and patterns of scientific research studies emerge.

Chapter 4

COGNITIVE AND TECHNICAL ASPECTS OF RESEARCH SUBJECTS

Before starting to read the chapter

Living creatures use their senses to collect data from their surroundings. Data collected via senses is processed in the brain. We, humans, can go beyond the knowledge that our senses provide us. We can collect large scale data from internal (senses) or external sources (e.g., documents and visual or audio records) in systematic ways (scientific methods) created by us. We can process data in many ways no other creature can do. Here we encounter our ultimate and unique advantage as *cognitive capacity.*

We need *cognitive skills* and *technical concepts* to conduct better research studies and better communicate our science. After this chapter, readers are anticipated to be able to recognize *different ways of reasoning.* Readers also know technical concepts that can help them to develop their research study in a better way. Readers are also anticipated to understand the *linguistic issues* related to science narration.

This chapter starts with a closer look at human cognition. The first part of this chapter presents content about cognition and related terms. Concepts related to cognition begin with *consciousness,* continue with *classification,* and finally end with *General System Theory* and *Chaos Theory.* Of course, the content presented in this chapter does not represent an exact and scientific classification of cognition-related terms. The terms and concepts shared under the subsection -Cognitive Dimension- are discussed mostly as definitions. The main purpose of writing this section is to provide the readers with an understanding of cognition and how we use it.

Final concepts of the cognitive dimension are *General System Theory* and *Chaos Theory,* which help us to comprehend the structure and motion of the physical and social world. This chapter gives a place for those theories due to they providing us with a consistent and integrated perspective of the world of facts, events, and objects.

The technical dimension of this chapter provides the concepts of *structure* and *integrity* in the context of research models and theories; *objectivity,* which is crucial for reaching truth; *semantics* and *use of language,* which are important for effective science communication.

HUMAN and TECHNICAL DIMENSIONS OF SCIENTIFIC RESEARCH

The Cognitive Dimension

Cognition is the use of conscious mental processes (Cambridge Dictionary, n.d.). Cognition can also be defined as the "ability to comprehend, mental act or process of knowing" (Bayne et al., 2019: 608). "According to constructivism, cognition is an active construction of reality by distinctions" (Matuszek, 2018: 204).

The initial sources of cognitive information are *memory* and the *senses,* which are processed by the brain. Second, there are external sources, such as *records* and *documents.* The act of reasoning lies at the core of the cognitive dimension of dealing with science. The most prominent, historically influential model of cognitive systematization is the Euclidean model of linear and deductive reasoning from basic axioms (Rescher, 2003).

The path followed in the act of thinking requires the use of logic. Cognitive activities such as *asking questions, analyzing, making inferences,* and *synthesizing* are at the core of the relationship between thinking activities and facts. The human mind always looks for patterns around us. For example, we tend to see human faces in objects around us. When we look at a mountain or cloud, we tend to perhaps see a human face. This phenomenon is called *face pareidolia.* It stems from our mind wanting to see familiar patterns. That phenomenon is conceptualized as a natural error of the face-detection system and reminds us of our mind's flaws (Wardle et al., 2020). Our mind works imperfectly and with biases such as selective perception, confirmation bias, etc. For example, selective perception stems from our needs, goals, expectations, and values (Shrum, 2008). Confirmation bias is to seek particular information to confirm a thought or proposal. It is an epistemically distorted way of seeking knowledge (Oswald and Grosjean, 2012; Peters, 2020). All of these examples of widespread flaws of the human mind show that our cognitive capacity has limits and flaws. These flaws require awareness because they can manifest themselves in various stages of the scientific research process, such as developing questions, creating research models, or making inferences. For example, cognitive processes provide presumptions, which are stipulated by scientific inquiry. Presumptions determine the "starter-set" of a study, and enable a starting point on whose basis further "inferential" reasoning may proceed (Rescher, 2003). This stage is prone to mistakes.

There are many concepts associated with the human mind. Recognizing the ways the mind works will also help you use it correctly. Let us take a look at the concepts related to cognition below.

Consciousness

Christof Koch (2018) describes consciousness as follows: "It is everything you experience. It is the tune stuck in your head, the sweetness of chocolate mousse, the throbbing pain of a toothache, the fierce love for your child and the bitter knowledge that eventually all feelings will end."

According to Cambridge Dictionary (n.d.) **consciousness** is the state of understanding and realizing something within one's mind. Consciousness is related to all mental activities and experiences of human beings (Lormand, 2005). Consciousness contains components such as cognitive activities, memory, interpretation etc. (Pereira and Ricke, 2009).

The functioning of consciousness is very important, especially in scientific activities, since the human mind has dark spots. Because selections and behaviors might not always be conscious. For example, in the process of creating a hypothesis, the researcher can formulate his own bias in a hypothesis unconsciously or, at best unintentionally (Fraenkel, Wallen, and Hyun, 2011). Therefore, researchers should recognize human behavior and mental processes correctly for the proper orientation of cognitive efforts in each stage of scientific research process.

Rationality

Rationality is the fact of being based on reason and logic rather than emotions (Oxford Dictionary, n.d.; Cambridge Dictionary, n.d.). Historically, rationality has been divided into two branches: (1) *practical rationality,* which concerns what we do; and (2) *theoretical* (or *cognitive*) *rationality,* which entails what we think. The activities of humans cannot be separated from reason. As a result, these two branches of reason form a whole (Rescher, 2003). Rationality plays a dominant role in gaining knowledge (Markie, 2005). Whereas, our rationality is bounded. *Bounded rationality* is a concept introduced by Herbert A. Simon (1955) in the field of economics and decision-making. It refers to the idea that individuals have cognitive limitations, such as limited information-processing capabilities and time constraints, which restrict their ability to make fully rational decisions. Thus, rationality is directly related to our knowledge-producing processes.

Thinking

Thinking is the activity of using the mind to consider something (Cambridge Dictionary, n.d.). Thinking is an individual activity, and many things, such as prejudices, beliefs, prior knowledge, can affect the results (a decision, a judgement on a subject, a solution for a problem etc.) reached through the act of thinking. Sometimes our subconscious can also play such tricks on us and direct us to certain conclusions in daily life.

You may have heard the media jargon phrase "echo chamber". It is a metaphorical expression that describes a situation in which individuals listen only to arguments that are close to their own worldview. A healthy thinking activity requires stepping out of the echo chamber, even in the daily pursuit of news and information.

There are many ways and types of thinking. I have placed thinking at the top hierarchically in this book. Let us take a look at several kinds of thinking and forms of mental activity.

Reasoning is an epistemic and mental activity. Reasoning is the process of reaching a conclusion from at least two thoughts, where one is treated as the proof of the other. Reasoning is a particular type of thinking done to draw conclusions or to prove something (Nickerson, 2011).

In the reasoning process, the most important input to reasoning is *knowledge*. Subsequently, the reasoning must be *logical*. Finally, all the factors mentioned above can also influence the conclusions reached through reasoning action. Thus, reasoning must be *objective*[1] that is, leaving personal prejudices and biases aside, is important in terms of reliable reasoning and reaching objective conclusions. We can say that these are leading standards of reasoning.

Analytical thinking is the act of thinking of something in detail to understand or explain it. Analytical thinking requires examining and analyzing the subject profoundly and in all aspects (Cambridge Dictionary, n.d.; Oxford Dictionary, n.d.; Radford, n.d.).

In order to reach a more comprehensive understanding of the act of thinking, let us now take a look at some other concepts related to the act of thinking.

Dialectic is a form of discussion or reasoning on a topic through a thesis and an antithesis (Oxford Dictionary, n.d.). Dialectical discussion or reasoning can be done to achieve a synthesis. Hegel defined dialectics as the movement of thought based on the triple framework of *thesis-antithesis-synthesis* and associated it with the movement of the universe, because Hegel saw the universe as a

---

[1] The objectivity principle is an important standard of making science and is not only related to the reasoning process. For more info, see Chapter 4.

"materialized idea." In another sense, according to Hegel, *thought* (mind) and matter are identical. Here, *dialectic* refers to the process of development of all thought and matter. This idea was later developed by Karl Marx and his proponents, and took the form of dialectical materialism, which briefly states that "*The flow of thought follows the flow of matter*" (Craig, 2005; Peng and Nisbett, 1999; Zwart, 2021).

A **metaphor** is a speech depiction in which a word or phrase that expresses a given type of element is used in place of another to suggest a *similarity* between them. A metaphor is a tool to develop creative and critical ideas (Martinich, 2005; Aubusson et al., 2006).

Three important actions are associated with metaphors: *comparisons, interactions,* and *speech.* First, each metaphor includes a comparison. In other words, every metaphor is an abbreviated analogy. Second, each metaphor includes a semantic interaction between an object (i.e., element, concept, fact, or entity) that is literally expressed, and a concept (target) that is figuratively based on the word. Finally, it is not the metaphorical words or sentences that matter, but their use in certain situations (Martinich, 2005). For example, the "*human body is a machine*" compares the human body to a machine. There is a semantic interaction between the *source* (the machine) and the *target* (the human body). The word *machine* is used figuratively (Aubusson et al., 2006; Martinich, 2005).

An **analogy** is a special type of *similarity*; it provides a simple way to explain abstract ideas in familiar terms. It is crucial to use analogies appropriately; otherwise, their benefits may disappear in unforeseen ways (Aubusson et al., 2006; Holyoak, 2012). Two elements (objects) are the same as the metaphor: the *source* and the *target.* In this way, analogy enables us to connect the unfamiliar to the familiar, the insensible or the abstract to the sensible (Park, Daston, and Galison, 1984). **Analogical reasoning** is based on systematic linking between the *source* and the *target.* If two facts share a common relationship between their elements, they are analogous. However, this link can sometimes go beyond the information given at the beginning and cause reasonable or incorrect generalizations about the target. In this regard, an analogy is a form of inducing reasoning. Typically, an analog, called a *source* or *base,* is more familiar or better understandable than the second analog, called a *target.* This asymmetry in the initial knowledge forms the basis of analog transfer, which uses the source to make inferences about the target (Holyoak, 2012).

The analogy has a linkage with imagination and models. The analogy can be used in scientific reasoning and building models. Models are elaborated analogies and more accessible to the senses. However, it has the potential to hinder as well as aid the understanding. Renaissance scientists and philosophers had various attitudes toward analogical use in science. For example, Galileo preferred the restricted use of analogy, whereas Descartes used analogy with more explicit forms of criteria, such as in intra-species comparisons. Bacon had a cautious attitude toward the use of analogies in science yet accepted analogy as a tool for scientific discoveries (Park, Daston, and Galison, 1984).

Making an analogy is also a method used in the world of science in scientific thought and in the determination of methodological principles. Making an analogy, as a form of reasoning, allows us to employ similarities and/or differences to evaluate existing experimental results while creating new scientific research fields. Reasoning by analogy supports the validity of a hypothesis by comparing the outcomes of an experiment with the findings of other experiments known to have similar features (Fisher, 2018).

Analogies are closely tied to metaphors and other forms of symbolic expression. Analogies are used in the history of science, mathematics, law (such as establishing a relationship between past cases and current ones), to influence political opinions, and to solve problems (Holyoak, 2012). For example, "the *human brain works like a computer*" is an analogy. The *computer* is the source, and the *human brain* is the target. The similarity is built in the way they *work*. Hence, the human brain and a computer can reciprocally enhance the understanding of the other.

Metaphors and analogies are not used consistently because, in the literature, there is no absolute distinction between a metaphor and an analogy. Even so, it seems that all analogies are metaphors, but not all metaphors are analogies (Aubusso, Harrison, and Titchie, 2006). Both metaphors and analogies are based on logic and abstract thought, because when we build a metaphoric or analogical statement, we transfer particular features of the source to the target.

According to the explanations above, an analogy is a more comprehensive activity of linking a source with a target than is a metaphor. An analogy is also much closer to scientific reasoning activities, allowing us to employ similarities and differences to better evaluate research findings.

Heraclitus argued that **conceptual thinking** is a form of the emergence of reality. Parmenides, on the other hand, asserted the priority of logic for conceptual thinking, and maintained that the senses are deceptive and that observation is a lower degree of input than a logical view (Stephens, Barton, and Haslett, 2009). Conceptual thinking is especially critical in the development of research models. Researchers use conceptual elements to design research models. Conceptual thinking helps researchers to gather ideas, principles, and elements related to something abstract, such as a social class, a system, or beauty.

Critical thinking is a purposeful, self-regulatory way of thinking; it necessitates relying on criteria, self-correction, and is sensitive to context (Hitchcock, 2024).

An example of critical thinking is the following: "*The type of reasoning known as -argumentum ad hominem- is a logical error, but it is so effective psychologically that it has* always been a standard form of popular reasoning" (Vander Nat, 2010: 5).

> **"Criticism is the lifeblood of all rational thought."**
> (Karl Popper 1974)

Abstract thinking entails using the mind to understand something via abstract tools. Mathematics is one of the best examples for abstract thinking (Mitchelmore and White, 2004). Analogies and metaphors are also examples of abstract thinking.

Lateral thinking is a form of problem-solving by thinking about a subject in a different and original way, rather than using traditional or expected methods (Cambridge Dictionary, n.d.).

Thinking may sometimes require somewhat *original/unique* approaches. For example, you can diversify your reasoning methods by consulting the opinions of others; especially people you know see the world differently than you do. Even if you do not share his worldview, you can get an opinion from someone whose knowledge or intelligence you trust. Sometimes the owner of this extraordinary mind may not be within your reach, and it may even be a writer, journalist, scientist or businessman that you admire. In this case, letting that person talk in your head can help you overcome your own thought patterns. Of course, in this case, you will need to know that person well enough to predict how his/her approach may differ.

We said that our way of thinking is influenced by many things. Additionally, different ways of thinking may have advantages but also disadvantages. For example, critical thinking can help you determine the deficiencies of a system, or sometimes, it can prevent you from producing a pragmatic solution. Or, when you focus on lateral thinking, you may neglect the factual reality and falter in finding a sufficient solution to a problem.

Sometimes, even the interpretations that the two genders bring to events may change according to their own priorities or even instinctively. Below, while describing the last scene of Episode 12, Season 3 of the popular television series *Friends* (1994-2004) by NBC, I am going to share a short dialogue in this scene with the words I remember.

> Ross has been with Rachel for a while and is happy. When Ross comes to the cafe where they always hang out, he tells Rachel that he is going to meet a stripper friend of his and that he and his son will accompany the stripper and her son in an activity. Rachel is a little surprised. When Ross notices this, he asks if there is a problem. Rachel says "*it is okay*", and then, as if to make the relationship safe, she approaches Ross and places

> a tight kiss on his lips. It is obvious that Ross was impressed by this. A goofy but happy expression appears on his face. Then Ross leaves the cafe. Rachel turns to Chandler and confidently says, "*He won't be able to get that kiss out of his mind for a few hours, huh*". There is a feeling of comfort on her face that comes from having sealed her man and tied him to her again. But Chandler replies to Rachel, "*Or maybe you got the guy horny and sent him to a stripper!*"

I think you can guess Rachel's facial expression. Here, we reason and interpret events differently -sometimes with an understanding specific to our gender- and we can make very different inferences about the same event. If even gender difference affects how we interpret the same event, I would like to draw attention to how political views, culture and one of its most important components, religion, can differentiate our perspective on events.

Analysis and Synthesis

Analysis is the process of breaking up a whole into parts or components (Barton and Haslett, 2007). For example, a *chemical analysis* of an asteroid reveals the components of the rock. *Data analysis* involves processing and interpreting collected data (Gay, Mills, and Airasian, 2011). **Synthesis** is the process of combining existing elements to create a coherent whole. Being able to analyze and synthesize requires reasoning and dialectical skills (Barton and Haslett, 2007). Synthesis is also used to point out the combination of a thesis and an anti-thesis as defined by Hegel- (Craig, 2005), which is sometimes suggested regarding a particular topic.

Context

As defined previously, **context** refers to all elements and/or properties that are related to each other via a fact, situation, event, process, or object. Contexts can appear in different forms, such as *interests, relatedness, situations,* and *sequences.* In this regard, concrete, measurable verbs such as *identify, describe, compare, sort,* and *classify* are included in scientific research as the context of research methods. Immeasurable, ambiguous verbs are useless for scientific research. Both quaantittive and qualitative studies should produce *identifiable, describable, comparable,* or *classifiable* results, and their results can be *sorted.*

When the context is expressed, the first thing that comes to mind is the cause-effect relationship between variables. However, in scientific research, it is not only cause-effect relationships that are sought. Sometimes, research can be shaped to examine the *correlations* between variables, or to estimate the *differences* (*heterogeneity*) between populations (by making a comparison), or to detect statistical *parameters* for a population. For example, to determine the success of undergraduate students attending a mathematics course, the parameters related to the scores they receive on the midterm exam are calculated as the arithmetic mean or the standard deviation. Here, the *variable* is the *score*, and the *context* is the *parameters*, which refer to a *situation*.

The purpose of research must be known in order to decide in which context the research will be conducted. This stage is critical. Sometimes, researchers may experience disappointment when they pursue logical contexts (especially by establishing a hypothesis based on a cause-effect relationship), gather data, and do not find a relationship between the variables. This situation may be tied to a lack of theoretical knowledge on the subject matter, or to the fact that the research is conducted in an erroneous context.[2] Of course, while making this inference, let us note that we excluded possible systematic errors that may occur during data collection or analysis.

Scientific research can produce results in terms of various contexts, such as:

→ Defining (one or more subjects);

→ Comparing (two or more subjects);

→ Sorting (two or more subjects);

→ Classifying (two or more subjects);

→ Calculating probability (expressed as a *percentage*), which refers to the possible future situation of a subject;

→ Calculating frequency (expressed in *units* and/or as a *percentage*), which refers to the present situation of the subject;

→ Calculating a correlation (which shows the *direction* and *strength* of the relationship or connections between variables, whether causal or not); and

→ Identifying the cause-effect relationship between independent and dependent variable(s).

Classifications

We have come across concepts such as *typology*, *topology*, and *taxonomy*. These terms have a general meaning that we can refer to as *classification* or *categorization*.

[2] There are various possibilities for contextual errors, such as linking variables to each other in a wrong way, and examining irrelevant features of the subject.

First, an *individual-general* connection is always established in logical operations, taking the *properties* into account. When we desire to take this process further, we come across distinctions such as genus and species. For example, a connection can be made between the two general concepts of *clarinet* and *saxophone.* When a link is established between two general concepts, it becomes a general-general connection. In other words, a general-general relationship is established between all saxophones, which corresponds to the concept of the saxophone as a class, and the concept of the clarinet, which is formed by all individual clarinets. These two general concepts come together via common features. For example, they are both wind instruments. Thanks to this common feature, we can create a new class by combining these two concepts. This upper class, which incorporates our previous general concepts, is based on the partial identity formed between the concepts. This partial identity, as mentioned above, is based on the common feature of these two concepts: They are both wind instruments (Jevons, 1893; Özlem, 1994).

When a relationship is formed between singular entities, a grouping, or *generality,* arises based on their common features. We might group a pen with other pens by associating it with other pens one by one in terms of their shared features (here, the dominant common feature is that all pens allow us to write). In short, the generalization process is a mental activity based on establishing identity rooted in common features. From this point, one general item concept is reached by starting from individual items. Singular entities can only become members of a class (e.g., genus-species[3]) if they share common features. The general concept is the universal answer to the question of *what* a thing is[4] (Özlem, 1994).

In logic, concepts including genus-species and the intension-extension[5] relationship are homogeneous concepts. On the other hand, concepts that do not have such a relationship are heterogeneous (i.e., of a separate type) concepts (Özlem, 1994). For example, while tablet computers and desktop computers are homogeneous, tablet computers and dustbins are heterogeneous concepts.

Thus, there is a logical basis for the classifications made, considering the *partial differences* and *partial identities* between the concepts. When two

---

[3] *Genus* indicates an upper class, while *species* denotes a lower class.

[4] We refer to it as a *term* in language

[5] *Intension* refers to the internal content of a concept, while *extension* means the range of applicability to particular objects. For instance, the intension of *ship,* as a substantive entity, is a "vehicle for conveyance on water," whereas its extension embraces things such as *cargo ships, passenger ships, battleships,* and *sailing ships* (Britannica).

concepts are linked and create a new structure, it takes on class-declarative names such as a *genus* or *species* (Özlem, 1994).

Species is a general concept that hierarchically falls under *genus* and is partially identical to *genus* due to some features. Let us remember that the saxophone and the clarinet (a genus) are wind instruments. It is always possible to reach more advanced, broader classes and concepts. As can be easily seen, the wind instrument class can be associated with the musical instrument class, which includes it by following the partial identity path in characteristics. Hence, wind instruments are associated with a higher class that enables one to produce sound for music. Therefore, the concepts in the wind instrument class form a subclass of the upper class (musical instruments). Likewise, musical instruments can be attributed to the instrument upper class. This relationship between genus and species leads us to the "rule of all" (*dictum de omni*). The rule states that a particular trait/feature (the most fundamental identical feature) found in the species cannot be excluded from the genus. In other words, a given feature attributed to a species covers all members. Here, the most basic feature is to be an instrument, which is also the highest *breed*. An instrument is the name given to objects produced for a specific purpose and can be used by human hands (Özlem, 1994).

Similarly, if we look from the top down this time, we can say that a feature that is not possible for the genus is not possible for the species. This relationship between genus and species leads us to the "rule of none" (*dictum de nullo*). Based on the definition of an instrument, there is a feature of being produced; that is, they do not exist in nature. Therefore, no instrument type and its underlying species can exhibit the feature of being found in nature (Özlem, 1994).

Categorizations or divisional labels are alternatives to this separation process, which can be referred to as *classification* (the most common name; Marradi, 1990). Let me mention from the beginning that I will use the terms *classification* and *categorization* in the same meaning as Bryant (1997) did, and as stated in the Oxford Dictionary (n.d.). Let us continue here by defining some of the concepts we are familiar with from everyday life.

Terms such as *class* and *category* refer to groups that emerge due to a classification or categorization process. The process of classification plays a crucial role in our conceptual lives. In daily life, we encounter many objects and events. When we categorize them, we combine various distinctive features (characteristics) and somehow treat them as *equivalent*. These similar features are grouped internally in the cognitive process, and information about these groups is selectively stored in memory (Bryant, 1997; Oxford Dictionary, n.d.).

The science of classification has its roots in ancient Greece. Aristotle believed that by nature, the world is divided into species and breeds; according to him, this is the natural order of things. Classification provides a wide variety of sub-functions, because classifying something as a member of a particular category allows people to recall the appropriate category information for the new instance. This information provides a basis for understanding or predicting actions, judgments, and decisions of encountered objects and events in the future (Bryant, 1997; Medin and Rips, 2005).

The purpose of class use in science is similar to the goal described above. However, the classification process, when dealing with science, is *logical* and *systematic*, and based on certain *rules*, *principles*, and *accumulative knowledge*. For this reason, classification processes in science can be called by different names. Let us start with the definition of *classification* first.

Classification is the process of separating a concept (fact, object, or event) at a certain level of generality into two or more narrow concepts with a lower level of generality. In principle, all aspects of the higher concept (apart from the aspect that is the subject of the classification) should be moved to sub-concepts. For example, if we classify dogs according to their color, the class of black dogs is characterized by all other features of being a dog except blackness. The essence of classification is the process of separating one class from other classes, which means drawing the boundary of a class. *Class diagrams*, *taxonomies*, *typologies*, or *topologies* emerge at the end of the classification process (Marradi, 1990). The basic elements of any classification consist of *theoretical commitments*, *basic units*, and *the criteria (variables)* used to put these basic units into a classification (Hull, 2005).

In the Encyclopedia Britannica (n.d.), **taxonomy** is broadly defined as the science of classification. The term is derived from the Greek *taxis* ("regulation") and *nomos* ("law"). We can trace its initial use to ancient Egypt and ancient Greece. Taxonomy is a classification created by conceptual studies based on a theory. It is widely used in botany and zoology, and organizes plant and animal species hierarchically as upper and lower groups (Britannica, n.d.; Winther, 2012). For example, in biology, living things are divided taxonomically into two realms: (1) animals and (2) plants. In the classifications made based on biological evolution, there are seven classes in the animal kingdom. Figure 4.1 presents the hierarchy in the animal kingdom.

Figure 4.1. Biological taxonomy

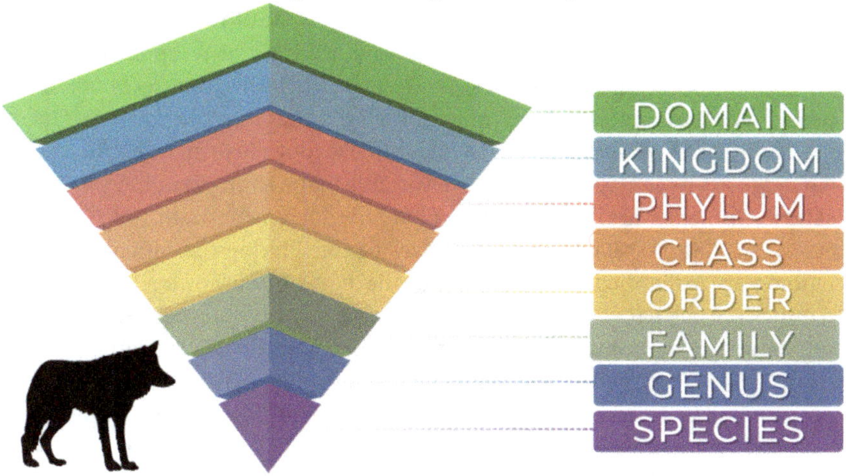

As mentioned above, there are taxonomic categories in the natural sciences. These categories that make up the tree of life are in hierarchical order. The realm is broken down into *domain, kingdom, phylum, class, order, family, genus,* and *species.* For example, humans are in the animal kingdom, in the Chordata phylum, in the class of vertebrates, in the mammalian order, in the primate family, from the genus of Hominidae, in the species Sapiens (Lunde, 2009). As can be understood from the example, to deal with taxonomies, certain concepts such as *rank* and *clade* are required (Marradi, 1990).

Aristotle realized that taxonomic classifications are not perfect. He noticed that even though species are very different from each other, cross-cutting classifications can occur. This is because no classification is perfect. For example, seals are both land and sea animals. Similarly, bats fall into both the bird and mammal classes. In sum, taxonomic classifications are not perfect because there are *intermediary types,* which makes it difficult to choose a certain class (Winther, 2012). For this reason, epistemological subjects are at the center of discussions in taxonomy (Hull, 2005).

Taxonomy and typology are used to classify certain approaches (such as strategic management or economics schools) in the social sciences. This is intended to create an analytical framework to better understand social issues.

Typology is also a science of classification, like taxonomy. Typologies make a major contribution to the creation of concepts and the determination of categorical variables (Collier, Laporte, and Seawright, 2008). **Typology** is the process of determining variable combinations to test concept types developed a priori. In other words, it is the process of developing dimensions that reveal concepts. A typology begins with a theory and creates a systematic basis for comparison. Taxonomies reveal patterns within variables. The main disadvantage of typologies, as a method of categorization, is that they are not exclusive (limited) in scope, but are usually based on arbitrary or temporary (ad hoc) criteria. In this regard, typologies are descriptive, rather than explanatory or predictive. Taxonomies differ from typologies in that they classify items based on empirically observable and measurable properties. In order to extract concepts expressed as types, discrete variable features are examined. With this aspect, typologies are categorical (academic.brooklyn.cuny.edu, n.d.; Kaufman et al., 2000; Smith, 2002). Also, the order of considering the fundamentals in typology is insignificant. For example, the types of authoritarian theocratic regimes are the same as those of theocratic authoritarian regimes. There are sequences in taxonomies in which the order in which the fundamentals are considered is important. This feature distinguishes typologies from taxonomies obtained from a successive basis. For example, the classification obtained by using the X feature to classify a biological genus and the Y feature to classify a species is by no means the same as the classification derived by considering the Y feature first and then the X feature (Marradi, 1990).

Typological distinctions are *qualitative* (e.g., living or inanimate beings, such as plants and animals) and aim to reveal categories (such as bird and fish) instead of grading, as in topologies. As a result, they define discrete variants. A topology refers to a degree of distinction, rather than categorizing species according to their contrasts; for example, color density, texture, and shape in drawings or paintings, or quantitative representations in mathematics. In this respect, topological parameters are based on continuous variability (e.g., shades of color on a spectrum). For this reason, it is more accurate to express distinctions in degrees. With this feature, topological studies focus on the *quantitative* differences of an object, system, or process. Other indicators of topological studies are *trends* and *continuity in transformation* (academic.brooklyn.cuny.edu). In sum, typologies classify types according to quality (e.g., warm colors, cold colors), while taxonomies produce ordered classifications according to sequence, hierarchical order, or quantity (e.g., ordering colors from warm to cold, or ordering colors according to their wavelength in physics). Finally, topologies create ratings (e.g., shades of yellow from lighter to darker along a gradient) and are quantitative in this regard.

The generation of scale categories, which I will explain in the next section, *Types of Measurements and Scales*, is also a type of typology. For example, categories that show the levels of education we see in surveys (ordinal scales) are typologies. Similarly, categories in which automobile types are defined (nominal scales) are typologies. Based on this last example, we can make a statement: Automobile typologies (categories) consist of sedans, sports cars, SUVs, et cetera. The categories that we define as sedans, sports cars, and SUVs represent a concept. In general, classification terminology is quite rich. Several examples of classifications are used in various fields, as follows (Hjørland and Nicolaisen, 2004):

→ **Biology:** Plant and animal taxonomies

→ **Linguistics:** In addition to the classification of languages, the classification of parts of speech (dimensions/segments; e.g., categorizing words as nouns, verbs, adverbs, etc.)

→ **Physical geography:** The classification of regions as tropical, subtropical, and temperate

→ **Cultural geography:** The classification of cities and countries

→ **Music:** The classification of music by genre and instrument by type (e.g., wind, string, percussion)

→ **Psychology:** The classification of abilities and mental events

→ **The social sciences:** Disciplines (categories) such as economics and sociology

As we mentioned before; science fundamentally *defines, classifies,* and *analyzes.* Thus, the explanation and examples of classification, which is one of the basic activities of science, are extensively covered in this book.

General System Theory and Chaos Theory

Underlying reasons for events usually exhibit *simplicity, order,* and *predictability,* unlike our universe, which contains *complex, chaotic,* and *random* features (Mitchell and Jolley, 2009).

When we think of the environment we live in, it is easy to infer that we are living in a universe consisting of systems. Our earth is part of our solar system, and our solar system is part of the universe we know as an astronomical system. Our earth contains geological and meteorological systems. We, humans, as living organisms, maintain our lives in an ecological system on earth. We are also living in social systems that can be described at different levels. We use mechanical and electronical devices as human-made systems. Therefore, it is very important to understand general system theory and to attain systemic thinking as researchers.

A **system** is a collection of interacting elements (parts/components/units) that exchange materials and/or energy. The behavior of a system as a whole can be distinctive from the behavior of its parts or elements (Marin, 1997; Higgins, 2002). Scientific research has analyzed basic properties of systems and discussed concepts such as *state, linearity, time-invariance, minimality, equilibrium, controllability,* and *observability* (Basile and Marro 1992).

General System Theory is concerned with classifying systems, determining the properties of systems, defining sub-units of systems, and finding solutions for the operational problems of systems. General system theory is a logico-mathematical field and aims to find common principles and laws that are valid and applicable to systems in general (Von Bertalanffy, 1950; Basile and Marro, 1992).

Another concept related to General System Theory is **Systems Theory,** in which "as an observer constructs a reality of self-reproducing psychic and social systems which also observe and construct realities by means of distinctions typical for each of them" (Matuszek, 2015: 208). As can be seen, while general system theory searches for patterns that can be applied to all systems; systems theory deals with more specific types of systems, such as psychic or social systems.

A natural, artificial, or social system may be simple or complex according to the number of its parts (elements) and static (stable) or dynamic (unstable) state according to its changefulness. The future action of simple systems is easier to predict under various conditions through the knowledge of relations between their limited numbers of parts (Higgins, 2002).

Systemic thinking involves features that systems exhibit. Mulej et al. (2004) suggested seven groups of system-related features: *interdependence* (relations, openness, interconnectedness), *complexity, attractors, emergence, synergy* (within a system) and *synthesis,* the *whole* (holism and the big picture), and *networking* (interaction and interplay; Drack and Schwarz, 2010). Therefore, systemic thinking is closely related to the **holistic view**, which involves interpreting facts, objects, or events and all aspects related to them (Marin, 1997). However, it should be noted that the aim of science is not only to study the whole in relation to the research subject it deals with; science can examine the parts and sub-components as well. For example, we need biological studies not only on the body or organ basis but also on the cellular level. For this reason, the science of biology, which studies living organisms, is accompanied by the field of molecular biology, which studies living things at the molecular level. Namely, biology can examine living creatures at multiple levels from at molecular level to the cell, organism, and population level.

Science tries to understand nature by reducing systems to their component parts (elements) (Higgins, 2002). We can imagine the *letter, cell,* and *atom* as the basic units of our universe. Each unit can be configured into *words, organs,* and *molecules,* which can be configured into *sentences, organisms,* and *objects.* Science is interested in the spectrum of possible states and the conditions of natural and social systems (Rescher, 2003).

Chaos Theory describes elements exhibiting behavior that is extremely sensitive to initial conditions that do not repeat, and yet are deterministic (Butz, 1995; Higgins, 2002). Chaos theory, developed out of work by Edward Lorenz on computing weather systems that is mostly known for its association with the *butterfly effect.* The random order of the weather system, displayed by Lorenz, looked like chaos. *Randomness* is the key property of the theory. However, a closer look reveals an underlying order. Thus, chaos theory is concerned with "*order in disorder*" (Butz, 1995); for example, making predictions about meteorological events.

Complex causal structures underlie social and natural phenomena. The complexity of social and/or natural phenomena makes it impossible to reveal all causes of events (Galbacs, 2020). This situation is reflected in the scientific perspective over time. Cohen, Manion, and Morrison (2007: 246) expressed this change in science as follows: "*Deterministic, linear and stable systems are replaced by 'dynamical,' changing, evolving systems and non-linear explanations of phenomena.*"

Random behaviors of the elements of a complex system have a certain probability. The complexity of a non-linear system is the subject of complexity theory, which is one step beyond chaos theory. Chaos theory is, as mentioned above, concerned with order in disorder, and concerns determining the future behaviors of a complex system validly, such as *meteorological events.* On the other hand, **Complexity Theory** is concerned with far more complex, non-linear systems, such as the *national economy* or the *stock market,* to which only probability calculations can be applied (Higgins, 2002). Complexity Theory is a form of non-linear systems theory that has attempted to capture what happens beyond what chaos theory describes (Higgins, 2002).

Chaos theory emphasizes the rarity of general laws in terms of scientific methodology, the need for huge databases, and the importance of examining exceptions in general patterns (Fraenkel, Wallen, and Hyun, 2011).

Cohen, Manion, and Morrison (2007: 33) explained the perspective of the two theories: *"Chaos and complexity theories argue against the linear, deterministic, patterned, universalizable, stable, atomized, modernistic, objective, mechanist, controlled, closed systems of law-like behavior which may be operating in the laboratory but which do not operate in the social world."*

In sum, there are two levels of chaotic systems. An example of the first level is meteorological events. The national economy or the stock market is an example of a second-level chaotic system. First-level chaotic systems are *deterministic* since they have variables and parameters specified by rules, and can be applied to elements through differential equations or other mathematical formulas. Hence, it is possible to determine future behaviors or states of elements of the system. On the other hand, second-level chaotic systems are *probabilistic* because they are far more complex and non-linear, and their future behaviors can only be reduced to probability calculations (Higgins, 2002).

Table 4.1 shows examples of systems according to their type and complexity. The term *system* as a concept can correspond to a social, natural, or technical phenomenon. While the human brain is a biological system, a computer corresponds to a technological system, and a social network corresponds to a social system. Roll the dice, decision-making, or social events are systems corresponding to several levels of complexity. System theory provides a thinking path when a researcher studies on features of a subject such as *relatedness*, *complexity*, and *networking*.

Table 4.1. Systems by complexity

	Type of System (Static or Dynamic)	
Level of Complexity	**Deterministic Systems**	**Probabilistic Systems**
Simple Sample / Low Complexity	→ Calculator (Linear and static) → Tape (Linear and static)	→ Roll the dice → Toss up
Complex Sample / Moderate Complexity	→ Computer (Linear and generally static) → Ballistic missile (Linear and generally static) → Traffic flow (Linear and generally static)	→ Stock exchange speculation
Very Complex Sample / High Complexity	→ Meteorological events (non-linear and dynamic) → AI -Artificial Intelligence- (non-linear and dynamic)	→ Human brain (cognitive processes) → Social events → National economy → Climate

Source: Basile and Marro, 1992; Chen and Stroup, 1993; Drack and Schwarz, 2010; Emanuel, 2013; Emanuel, Ravela, Vivant, and Risi, 2006; Higgins, 2002; LeCun, Bengio, and Hinton, 2015; Rivera, 2020.

In Table 4.1, for example, meteorology is designated as deterministic owing to its capacity for providing relatively precise short-term weather forecasts. This deterministic quality is rooted in meteorology's focus on immediate atmospheric conditions and phenomena, allowing for a degree of accuracy in forecasting within shorter timeframes. Conversely, climatology, concerned with longer-term climate patterns and trends, is characterized as inherently more probabilistic. This probabilistic nature arises from the complexity and variability of factors influencing climate dynamics over extended temporal scales, leading to greater uncertainty in long-range prognostications (Emanuel, 2013; Emanuel, Ravela, Vivant, and Risi, 2006).

The Technical Dimension

Another topic that can be handled under the human dimension is the technical dimension in the context of science, the philosophy of science, and research methodology.

Technical topics, examined in this section, cover epistemology, semantics, preferences of keywords, the use of language, structure and integrity, and objectivity in science.

Objectivity and Subjectivity

Objectivity is the commitment to reality without being influenced by personal beliefs or feelings. In other words, objectivity is the commitment to the existence of the real world, which is independent of, and external to, the human mind and, thus, its experience or cognition (Bryant, 1997; Buhringer and Sassen, 2010; McMillan, 2015). Objectivity is difficult to apply to religious practices and cultural interpretations. Although a paradox may arise here in terms of our search for scientific knowledge, we must establish a balance that takes into account the differences in understanding between people with our search for scientific objectivity (Kogler, 2007).

Achieving objectivity is easy in some cases, but difficult in others. When making a judgment about an object's shape, for example, when we say, "this table is square," it is easier to make an objective judgment. However, making objective judgments on ethical subjects is more difficult. For example, when we say, "murder is wrong," this proposition will be seen as more controversial (Miller, 2005) because new questions arise in the mind, such as -*Under which conditions?*- and -*Even while trying to protect ourselves?*-

Objectivity is an important quality indicator of research studies. In outline, the researcher, sampling method, and measurement tool are anticipated to be objective. In general, objectivity should be considered at every stage of the research process. It starts from the sample selection stage and continues toward collecting data, performing analysis[6], and carrying out the conclusion (McMillan, 2015; Outhwaite and Turner, 2007).

---

[6] *Validity* and *reliability* (see Validity and Reliability in Chapter 7) are a crucial part of methodological objectivity, as they indicate the compatibility between the scale and the measured population (or subject).

Achieving this standard can sometimes be challenging. To maintain objectivity, researchers should employ available theories both in the sampling process and about the subject matter. Theories also serve to uphold certain standards by directing the sampling process and guiding the selection of measurement tools (McMillan, 2015; Outhwaite and Turner, 2007).

Objectivity in the sampling process is measured by the ability of the sample to represent the universe. Sampling techniques are guiding for researchers, especially in field measurement studies. On the other hand, in case studies, especially in studies where more than one case will be selected, it would be beneficial if the selected cases have similar patterns that will provide information about the subject under study or represent a similar level. Sometimes, in cases that require comparison, case examples can be specifically selected from different designs. All of this relates to the purpose of the research, and sampling objectivity does not idealize a single approach (Outhwaite and Turner, 2007). For example, in a comparative socio-political study conducted with two individuals from radical movements of two different ethnic (or cultural) communities in a study examining radical (fundamentalist) political movements, sampling objectivity requires that the samples (in this case, individuals) selected from both communities represent the most radical movements. If two samples of individuals representing one of the societies are representatives of a less radical movement than the other, the objectivity of the sampling will be violated.

A measurement tool is objective when different people obtain the same value using it (McMillan, 2015; Outhwaite and Turner, 2007). Even though objective tools are used in the research study, this does not always guarantee objective conclusions. Because researchers are anticipated to make inferences of objective truths. We come across the researcher's worldview as another subject of objectivity. A worldview is a person's subjective representation of external reality. According to Johnson (1999), a worldview determines how a person understands the world, and, as a result, affects his/her behavior deeply (Stephans, Barton, and Haslett, 2009). According to Polanyi (1958), personal judgments become part of science, especially in scientific studies carried out by a single person. Polanyi expressed this effect on a scientific outcome as a personal coefficient (Kochan, 2013).

In sum, researchers should aim to minimize the influence bias arising mainly from sampling, the measurement tools they use, or their worldview to avoid subjectivity and to produce a study that reflects the truth (Bryant, 1997; McMillan, 2015; Outhwaite and Turner, 2007). On the other hand, in some cases, subjects such as *culture*, or even some disciplines such as *politics* and *history*, as we mentioned before, may permit and include subjectivity partially. It is also accepted that in some disciplines, local beliefs, value judgments, and culture can have an impact on research studies, such as those conducted in *sociology, educational,* or *business administration* studies, and yet there may not be a problem of objectivity.

Semantics

Semantics is the systematic study of meaning. Current studies in this field are based on the work of both philosophers and linguists. Semantics seeks to answer questions about how meaning should adhere to truth and understanding. In addition, numerous semantic problems related to certain linguistic structures are among the discussion topics of philosophy (Crimmins, 2005).

Logic and language are intertwined phenomena. The correct functioning of logic depends on the choice and syntax of words. In this regard, we can accept semantics as an important input to logic.

Themes related to semantics, such as the use of *ideas, terms, meanings,* and *concepts,* should receive attention in a scientific sense. Unfortunately, semantics cannot guarantee representing all elements of the idea of a truth. Further, ideas, terms, meanings, and concepts might not be equivalent for different persons (Henderson, 2007). In this respect, scientific research studies also contribute to the production of meaning.

There are different approaches to semantics in the philosophy of science. For example, logical empiricism is characterized by the *meaning.* This approach attempts to define theoretical terms by virtue of the meaning of purely observational terms (Neuber, 2014).

Intensive discussions were held on propositional statements in the 20th century. Some philosophers asserted a distinction between *analytical statements* and *synthetic statements.* Willard V. O. Quine was one of the authors at the center of discussions (Meeker, 2011) that rejected the distinction and examined the logical bases of statements in the context of *meaning* and *naming* (Quine, 1951). Quine indicates meaning and naming may not be identical. For example, the Evening Star is the planet Venus, which is also the Morning Star. As can be seen, there are two naming referring to the only meaning. All those discussions show us the importance of logical basis and semantics of pursuing science.

On the other hand, it is crucial to know and use precisely the corresponding meanings of the terms and concepts of a field from a scientific point of view. For this reason, the definitions that meet the terms and concepts should be correct (within the framework accepted in the literature) and include all elements (components) available, thus representing the concept.

Philosophers prefer to seek the meaning of a particular word not in isolation, but in the context of a proposition (Marvin, 2011). Similarly, in scientific research, a concept or term should be handled and limited within the context of its own field and the proposition made. Concepts such as polysemy, the uncertainty of the meaning of a term or word, and using closely synonymous words interchangeably might require more attention (Henderson, 2007).

Terms, Terminology, Keywords, and Concepts

Each scientific and professional field has its own technical vocabulary. Each of these words is called a *term.* A set (or body) of terms specific to a field is called *terminology* (Oxford Dictionary, n.d.), for example, medical or legal terminology.

A keyword is essentially a term related to the studied subject. In scientific research, it is expected that the keywords used in formulating the research question will be compatible with the field's terminology. In addition, selected keywords/terms should have a context with the research problem. Therefore, it is important to reveal the *constitutive definitions* of the keywords clearly to carry out the research according to the purpose. There are also *operational definitions* that are vital for measuring keywords (Fraenkel, Wallen, and Hyun, 2011).

It is possible to choose the wrong terms, especially if the terminology of the relevant literature is not sufficiently known. For this reason, while conducting a scientific study, the main works of the related literature should be read carefully. Thus, the researcher obtains his/her theoretical background, along with the terminological readiness for conducting an academic study.

The keywords in the research question must be correctly and clearly defined, both in terms of their constitutive and operational meanings, without any ambiguity. The reason for this is that ambiguity in the meaning will prevent the correct execution of the research. This situation may misdirect the research and prevent it from attaining its purpose. Let us examine two examples.

> Motivation is a word/term that we often hear and use in everyday life. The importance of motivation is also accepted in professional life. *Student motivation* relates to the educational sciences, while *employee motivation* is related to business life. In educational life, teachers want to work with students who are inspired to learn, and managers want to work with employees who are driven to work in business life. Given its widespread position, the subject of motivation is frequently examined in the social sciences.
>
> In a research study on this subject, a researcher may identify a lack of motivation as a problem and may want to identify "*the factors that reduce motivation.*" The purpose of such research is to eliminate the factors that lower motivation and thereby increase students' motivation.

As an alternative, a researcher can focus on identifying "*the tools that increase motivation.*" In other words, motivation-enhancing incentive tools can be explored.

Regardless of the purpose, we need to start research by defining the term (constitutive meaning). **Motivation** is a person's desire to achieve a goal and to take action to do so (O'Connor and Carol, 1995; Robbins and Coulter, 2012). The definition of motivation reflects its *constitutive meaning*. However, this definition does not give us a clue about its *operational meaning*. Terms are generally technical concepts, and a term used in a particular area corresponds to a specific operational meaning within that area or refers to a specific application. For this reason, the researcher should also determine the *operational meaning*; that is, the actions and operations that characterize motivation (Fraenkel, Wallen, and Hyun, 2011). Therefore, answers to the following questions are sought: "*What is the operational meaning of motivation?*" and "*How do we measure motivation?*"

The operational meaning of motivation refers to *inner forces* (Watkins, n.d.). Motivation can be measured behaviorally. For example, an *increase in performance* is an indicator of motivation; this indicator can be included in the operational sense of motivation. The reason we made this distinction is to highlight the difference between the constitutive and operational meanings of keywords.

In our explanation above, the reason we said that a term should be handled in accordance with the field it belongs to is because of the differences in operational meaning. In terms of student motivation, students' *performance* (i.e., *exam results*) can be accepted as a benchmark of motivation. On the other hand, performance criteria in business life are gauged by *productivity* (or task-oriented outputs). Therefore, performance is a conceptually common yardstick of motivation. As can be seen, the operational meaning varies according to the field.

Let us take this mental exercise one step further in the context of motivation and show the further limitations of the conceptual skills required for scientific research.

We said that we can take *performance* as a benchmark to measure motivation. Let us imagine a student. She/he devotes more time to studying outside the classroom for a music lesson than other students. Hence, it seems we have a motivated student. However, she/he does not seem very successful while performing with his/her instrument in the classroom. For this reason, she/he receives a score that allows her/him to only narrowly pass the class. Consider that we have "*a motivated student who performs poorly.*" Thus, we might not find a relationship between motivation and performance in all circumstances.

Performance might not be a unique indicator of motivation! At this point, we encounter *talent*. The same is true in the business world. If an employee with high productivity (a seemingly driven employee) is not able to achieve the quality expected, perhaps she/he does not lack the motivation to work. Let us end this exercise here.

The term *concept* isn't easy to define exactly. Additionally, a concept may evoke multiple meanings. Concepts are often thought to be the meanings of words and are also related to our conceptualization of the world or our beliefs about the world. We conceptualize our world and constitute classes. For example, mammals are a biological class and correspond to a concept in our mind. According to psychologists, concepts correspond to internal representations of things. According to logicians, a concept is a set and functions of real and possible objects (Rey, 2005).

Use of Language

We can only express our thoughts—which are the products of thinking activities—through language (Jevons, 1893). Since language is a social phenomenon and has a technical aspect, it is included in science. Further, we convey concepts, theories, and findings through language. Let us briefly check on the importance of *rhetoric, the choice of the right words,* and the *use of grammar* with regard to science (Katz, 2009; Lynch, 2007). I should mention that there are widespread debates in the scientific and philosophical world on scientific language. Ludwig Wittgenstein (1889-1951), among the pioneers, worked on the philosophy of language. Rudolf Carnap worked on scientific language. Both philosophers emphasized the potential problems of language in both science and philosophy. Here, I present some practical suggestions for the use of scientific language without going into the discussions of how much the *concepts* and *meaning* represent reality (see Positivism and Post-Positivism in Chapter 2).

The language of scientific literature should be *precise* and *plain*. Researchers should avoid making vague statements. All explanations, inferences, and discussions should be precise, as the essential characteristic of scientific writing is clarity. The purpose of using precise, plain language in science is to eliminate the possibility of confusion. Vague phrases can also cause the other party (the receiver in the communication process) to misunderstand the explanations (Gay, Mills, and Airasian, 2011; Katz, 2009; Mitchell and Jolley, 2009).

Using short sentences is a way to ensure the clear use of language in scientific writing (Gay, Mills, and Airasian, 2011; Katz, 2009). As an explanation gets longer, the possibility of grammatical errors increases, and so does the possibility of the receiver misunderstanding.

Word preference is an important subject in scientific writing. Researchers should avoid using assertive, unclear, or redundant words. Selected words and phrases should also be objective because scientific writing relies on logic—not emotions (Katz, 2009). It is critical to make sure to use *consistent terms* in scientific research. A researcher should also avoid using synonyms because they might cause the reader to misunderstand the ideas in the research report (Creswell and Creswell, 2017).

The use of grammar in scientific research is another central issue. Adjectives, adverbs, conjunctions, and prepositions can cause semantic shifts when not used properly. For example, "because" is used to report cause and effect since it refers to a reason that caused a result. This is the correct use. "Though" and "although" are conjunctions that report contrasts. However, they are often misused (Gay, Mills, and Airasian, 2011; Katz, 2009; Weathington, Cunningham, and Pittenger, 2012). Mistakes can also be made in the use of adverbs, adjectives, and adverbs (Katz, 2009). For example, some adjectives may seem near-synonymous and researchers might not see any problem using them interchangeably. In fact, both adjectives might differ from each other in terms of degree.

Spelling is another problem in writing a scientific paper. Researchers might use a computer-based spell-checker to avoid misspelled words (Katz, 2009).

In the scientific literature, researchers may intend for findings to be presented in a striking way. On this point, the last important concept in the academic literature is **rhetoric,** or the theory and practice of producing arguments; it is defined as the art of speech. Of course, it is vital to transfer findings accurately and completely. However, the effort to make results appear striking should not result in exaggerating them! There is a fine line between being practical and exaggerating. Researchers who study social and political issues tend to use rhetoric, but they should avoid using biased or manipulative rhetoric to build their conclusions (Lynch, 2007).

Structure and Integrity

Scientific writing has two main pillars—structure and integrity—that form the theoretical framework and shape the implementation of research. Within a research study, the *conceptual framework, research hypotheses* (or *questions*), and *research design* must complement each other. Hence, these two basic legs should be correctly structured, and logical integrity should be provided between the internal elements of these two legs.

A research study must be integrated into the theory, which is the existing knowledge accumulation, within the field (Kogler, 2007). A research study must provide consistency within itself to ensure integration in the conceptual framework, research design, and connection of findings to theory.

Coherence is a critical feature of the structure and integrity of scientific research. Ideas are tied together and flow logically. Terms, theoretical elements, and findings should be linked to each other in a coherent way. To ensure the use of consistent terms in a study, variable names in the title, the statement of purpose, and the research questions should always be kept in mind, and the researcher should follow these elements in each phase of the research to ensure integrity (Creswell and Creswell, 2017; Patton, 2014).

More attention may be required to achieve integrity, especially in case studies. On the other hand, researchers should not overlook the particularity of each case while examining common patterns and themes (Patton, 2014).

Relationship between Theory and Practice

A theory involves a set of constructs and their interrelationships. More explicitly, a theory has two components; *concepts* or *categories*. Those components that make up the theory are considered to be in a relationship. A theory possesses bounded explanatory and predictive power. Theories derive insights from the specifics of time and place. More specifically, inspirational stimuli are often used for creativity to promote novel ideas. Here, *stimuli* and *ideas* are constructs in a relationship that can be abstractly described for most creative design situations, irrespective of when or where they take place. Stimuli help to generate ideas. *Stimuli* and *ideas*, as variables, and their relationship hold common meanings in the context of design creativity, pointing to an underlying body of knowledge. Theories are central to the scientific endeavor in applied fields such as design. Thus, "generic" tools, methods, or processes can be developed for field applications (Cash, 2020; Maxwell, 1992).

There is a strong relationship between the scientific and practical worlds. The scientific world tries to comprehend the reality of the world and its problems. Hence, scientific studies develop theories, and professionals try to apply these theory-based solutions. For example, the concept of risk indicates situations that include the probability of loss and uncertainty. Insurance systems calculate risks for particular situations (e.g., car accidents) and offer practical options to avoid possible losses. In this case, *Probability Theory* helps insurers for practical applications. Similarly, *Fuzzy Set Theory* helps insurance practices (Shang, 2006).

Other examples include *Game Theory* and *Evolutionary Game Theory*, which are used in strategic decision-making processes in the business and political realms. Similarly, *Network Theory* has applications in many social and economic fields (Golbeck, 2013; Snidal, 1985;).

CONCLUDING REMARKS

Diverse cognitive skills and technical qualifications related to scientific research can make a great contribution to a researcher who is willing to accomplish a qualified research study. Some concepts and topics discussed in this chapter are anticipated to contribute modeling of research or processing of research findings, and some to contribute to science communication.

SUMMARY OF CHAPTER 4

The human mind distinguishes itself from the mental capacities of other creatures through its unique capabilities. Our activities of reasoning and pursuing science are dependent on our unique *cognitive capabilities.* We use collected and processed data to conclude, interpret, or make inferences about a subject. We use cognitive skills such as *rationality, reasoning* to conclude, interpret, or infer from processed data. Various cognitive skills can be used in various ways depending on the subject and purpose of the research study. For example, some studies need more conceptual thinking while others need critical thinking; some studies need making classification of variants while others need to make predictions.

Another cognition-related concept is *context,* which refers to the relatedness of elements and/or properties. We use this cognitive skill in a particular context(s) to build a relatedness of elements and/or properties of a fact, situation, event, process, or object. Building a correct context between variables is crucial for a research study to be settled on a logical base.

A *classification* is a useful tool for a better understanding of classes of objects or living creatures. Classification helps us conceptualize and group converging units. It also helps us understand transition criteria between classes.

General System Theory and *Chaos Theory* are related to our physical world and provide us with a useful perspective of the world. The physical world is chaotic and consists of systems and sub-systems related to each other that are changing in time. Those theories support our cognitive processes through their conceptualization of the physical world and help us to comprehend the relationship of the whole to the part, of one system to another

Semantics, the *use of correct terminology,* and the efficient *use of language* are crucial concepts and topics for effective science communication.

Crucial aspects of a research model are building the correct *structure* and providing the *integrity of patterns.* Furthermore, conducting reliable and valid analysis requires the objectivity of the researcher.

Chapter 5

MAIN CONCEPTS RELATED TO SCIENTIFIC RESEARCH

Before starting to read the chapter

When we decide to be a professional in a particular area, initially we start assimilating concepts related to the area. Pursuing science as a professional activity is not free from this need. This chapter prepares readers by introducing important concepts related to the actual execution of the scientific research activities. Readers will find comprehensive information about the scientific concepts such as *scientific method, the variable, data collection methods, population and sample, parameters and statistics, paradigms and modeling,* etc.

The other important concepts to follow are; *experimental* and *control groups, measurement* and *scale,* the distinction of the *unit of measurement, and level of measurement, the unit of analysis* and *level of analysis.* The last topic of this chapter is p*aradigms* and *modeling.* As can be seen, there are many concepts relevant to scientific research. Unless those concepts are comprehended well, it would not be easy to implement a scientific research study correctly.

This chapter is also structured to show differences between related or close concepts, as pointed out above. For example, a distinction between concepts such as data, information, and knowledge might not seem so important at first glance, but we should keep in mind that it is functional and necessary in many situations to distinguish between close concepts and notions to prevent misconceptions.

SCIENTIFIC RESEARCH and IMPORTANT CONCEPTS

Scientific research is the process of systematically collecting, analyzing, interpreting, evaluating, and reporting data to seek reliable solutions to problems or questions. In short, scientific research is a systematic process of data collection and analysis. I will include some important concepts and terms regarding scientific research methodology under the following titles.

Entities such as *firms, public institutions, foundations,* or *organizations* conduct research. The purpose of doing research is critical. Research can be carried out for economic or commercial reasons, such as to make technical progress or to better understand an object, process, phenomenon, or social issue. Not all research activities may be aimed at pursuing science, but it is imperative to use scientific methods in order for the results to be accurate and objective.

Firms can conduct research to determine market trends, and their position in the market, or to monitor developments in the sectors or activities of their competitors. Public institutions, non-profit institutes, and foundations can carry out studies to determine certain facts and situations in accordance with their own purposes. Finally, people can perform research. A trader can conduct industry-specific, region-specific, or general market research to further his/her personal business activities. Further, there are people who carry out research while pursuing postgraduate studies as part of their academic careers.

Scientific research implies both financial resources (money, equipment, etc.) and human resources. For example, a study in a laboratory environment may require technical equipment. Social research may require data collection through a questionnaire or interview. Research studies are also restricted by the limits of these resources. Since repeating an experiment using (perhaps rented) expensive laboratory equipment (with limited time) and expensive consumables will create extra costs, the experiment must be done correctly in one go. Similarly, social research conducted via interviews should be designed appropriately, and the right sample should be chosen from the very beginning, considering the costs of reaching the sample audience, again, in terms of material and human resources. In short, limited research resources must be used effectively so as not to require repetition.

The most important difference between the natural and social sciences is that humans, as the subjects of the social sciences, have the ability to think freely and act. While this reduces the predictability of social events, it also causes the factors that trigger social events to show more complex features and intricate relationships. Therefore, social issues can only be reduced to statistical estimates and probability calculations, if that is the goal of the research.

Scientific Method

The basic content of the scientific method consists of a set of general rules, criteria, and instructions. However, the suitability of a scientific method may be ambiguous under certain conditions; when applied in new research fields or interdisciplinary fields, the scientific method should take a form suitable to the nature of the new field (Bohm, 1961).

The notion of a method is in conflict with the epistemological notions of *openness* and *flexibility*. Epistemology, as a guide for methodology, presents a dilemma that can only be resolved by redefining "method" (Dahlberg, 1995).

Scientific method depends on the systematized procedures required to identify the most economical (e.g., simple, general, straightforward) solution that solves a given problem. Scientific method necessitates simplicity and systematicity. The *principle of simplicity* is one of complexity management, because simplicity allows the researcher to move onwards from the simplest (least complex) available solution to introduce further complexities (Rescher, 2003).

The Variable

One concept that many young researchers have difficulty understanding is the variable, which can be confusing because of its different types. Starting from the definition, it is easier to make inferences by looking at the meaning of the word itself, as in many concepts. You might think that the concept of the variable should be about something that varies.

Each element and feature of a fact, person, object, or event studied in scientific research is called a **variable**. In other words, a variable is an indicator of a state, characteristic, or element of a population that can be qualitatively or quantitatively defined. Variables are non-constant properties; that is, they are *varying*. A variable can correspond to a *change* that is individually experienced (i.e., a fact, person, object, or event), as well as variability within a population. Let us express it with a simple example. *Age* is a variable. As time passes, our age changes. This refers to the variability (as a unit of analysis) for a person. At the same time, the ages of individuals within a certain group of people will also be different (i.e., they will vary) from each other. This refers to variability (as a unit of analysis) for a group.

Since the definition emphasizes variability, it also indicates the *heterogeneous* features of the units. These distinctive features of a population may be age, weight, gender, educational status, income, level of motivation, membership status, volume, color (for an object), organizational structure, communication model, a company's understanding of management, the industry in which a company operates, or a company's position in the market. In order to better understand what a variable is, it is useful to clarify the concept of *constant*. Members of a cluster (that may consist of objects or persons) are seen as *identical* when they do not differ from one another. In this context, since there is no variability, no variable can be examined. Therefore, these properties are expressed as constants. In short, a constant is any characteristic or attribute that is the same for all members of a given group. If all students in a class have blue eyes, eye color is a constant (not variable). Some of the characteristics of the subjects discussed in a study are constant, and some are variable (Baur, 2011; Fraenkel, Wallen, and Hyun, 2011).

Remember the various survey questions you have come across! In these surveys, you may have been asked about your age, gender, education, income level, or how many children or siblings you have. These are all variables. Suppose you are asked about your eye color (which is also a variable) for a statistical study of eye color ratios in your country's population. Be careful: They ask you about your eye color, but they do not ask how many eyes you have, because the number of eyes is a constant—not a variable.[1]

A variable refers to the *distinctive* feature (property) of a studied unit. The units studied vary depending on the area and subject. For example, living creatures such as birds, fishes, and trees can be the subjects of a study in the natural sciences. Human issues or events such as a person, group, organization, department, country, culture, conflict, and immigration can be the subjects of a study. Objects or entities such as newspapers, software, movies, devices, and vehicles can be the subjects of a study in the social sciences or applied science (Baur and Lamnek, 2016; Fraenkel, Wallen, and Hyun, 2011).

Variables are used to classify cases (units) concerning the category they belong to. In the social sciences, cases (units) can be persons, departments, or organizations as members of a wider group, society, organization, network, or social system. For example, the feature of being "female" can be attributed to the *gender* variable, while "American" can be attributed to the *nationality* variable. Following the same logic, the movie *Titanic* can be referred to the category "1997" as a *production year* variable, while "PG-13" is a *rating* variable, and "11" is the *number of Oscars won* variable (Baur, 2011).

Variables are strongly related to research question(s) and hypothesis in scientific research. Because of this, researchers should pay attention to binding research questions or hypotheses that are directly related to the variables studied (Creswell and Creswell, 2017). Variables can vary according to their *features* and *roles* in a study. I explain the types of variables in detail under the following titles.

Types of Variables

As previously mentioned, variables can be classified according to their features and roles in a study. A broad explanation of the concept of a variable is given below.

---

[1] In this statement, of course, we exclude congenital anatomical defects and eye loss due to several reasons.

Variable Types According to their Role in the Research Model

The most common types of variables, according to their roles in a study, are as follows:

- ⇨ Dependent variable
- ⇨ Independent variable
- ⇨ Control variable
- ⇨ Moderator (regulatory) variable, and
- ⇨ Mediator (intervening) variable

Let us examine these one by one.

Research can be done using a single variable or multiple variables. Research involving more than one variable generally includes studies of cause-effect relationships; in such studies, the relationship between the two types of variables is investigated. According to their role in the research model, these variables are characterized as *dependent* and *independent.* Simply stated, in a cause-effect relationship, the independent variable expresses *the reason* (cause), and the dependent variable expresses *the result* (effect) (Creswell and Creswell, 2017).

The **dependent variable** is the variable where the influence (effect) of the independent variable is observed. In other words, it is the variable that depends on the independent variable (Creswell and Creswell, 2017). This element is often uncontrollable. However, control can be achieved with the change made at the level of the independent variable. For example, in a study, the researcher can decide that "one glass of red wine consumed daily in a particular period" (e.g., eight weeks) is the independent (or manipulated) variable, while the dependent variable is cardiac or vascular health.

The **independent variable** is the variable whose effect is observed on the dependent variable. In other words, it is the variable that is manipulated in an experiment, and is thus independent of all other effects. The independent variable is also known as the *manipulated variable,* the *experimental* variable, or the *treatment variable* (Creswell and Creswell, 2017). The independent variable can also be seen as a factor in the research model. This element can often be controlled.

Other examples include "by increasing advertising spending, the sales volume will increase" and are presented in Figure 5.1. The arrow indicates the direction of the effect, which moves from the independent variable (e.g., income) towards the dependent variable (e.g., consumption habits).

Figure 5.1. Direction of the effect from the independent variable towards the dependent variable

Independent Variable (X)		Dependent Variable (Y)
Intelligence (quantitative)		School achievement (categorical or quantitative)
Education level (categorical)		Income (quantitative)
Income (quantitative)	➔	Consumption habits (categorical)
Advertising expenditures (quantitative)		Sales revenues (quantitative)
Gender (categorical)		Social reaction (categorical)
Population growth rate (quantitative)		Energy consumption (quantitative)
Exam preparation time (quantitative)		Exam score (quantitative)

The findings of studies examining the effect of an independent variable on a dependent variable can be shown in graphics, which can make it easier to present the direction and strength of the relationship.

A third type of variable in studies (in the model) is the **control variable**, which is kept *constant* in research. Generally, demographic characteristics (variables) describe individuals or the size, age, ownership, export density, and management approach of companies. For example, in a study of the relationship between fatigue (the independent variable) and learning (the dependent variable), a certain weekly (standard) working time can be taken as a control variable to obtain a healthy (accurate) result. In this context, while students who comply with the weekly working standard are included in the measurement, those who fall outside the standard are not included in the measurement. Thus, thanks to the control variable, the effect of the independent variable on the dependent variable can be observed/measured correctly, which makes the findings more accurate.

Another variable is the **predictor variable** (also called the *antecedent variable),* which is used to predict an outcome of interest in surveys. The predictor variable is similar to the independent variable, and is hypothesized to affect outcomes, but differs in that the researcher is not able to systematically manipulate it, as the predictor variable may or may not be feasible. The **outcome variable** (or the *criterion* or *response variable*) is considered the outcome or result of the predictor variable(s) in surveys (Creswell and Creswell, 2017).

Two more variables appear when the dependent and independent variables of a causal relationship are influenced by one or more additional variables. The first one, the **moderator variable**, is the predictor variable that affects the direction and/or strength of the relationship between the independent and dependent variables, or between the predictor and outcome variables. In other words, in a causal relationship, when the moderator variable coexists with another independent variable, it can *strengthen, weaken, buffer*, or *reverse* the effect of the independent variable on the dependent (criterion or outcome) variable through its effect on the relationship with the independent variable. The effect of the moderator variable on the dependent variable is observed indirectly (Creswell and Creswell, 2017; Rosenstein, 2019). More specifically, the moderator variable expresses a strong contingent effect in the relationship between the independent and dependent variables. Thus, the statistical relationship (e.g., correlation) between two variables changes as a function of the moderator variable (Sekaran, 2003: 91; Weathington, Cunningham, and Pittenger, 2012). Figure 5.2 shows a research model with a moderator variable along with independent variable A and dependent variable B.

Figure 5.2. Relationship model: Example of a moderator variable

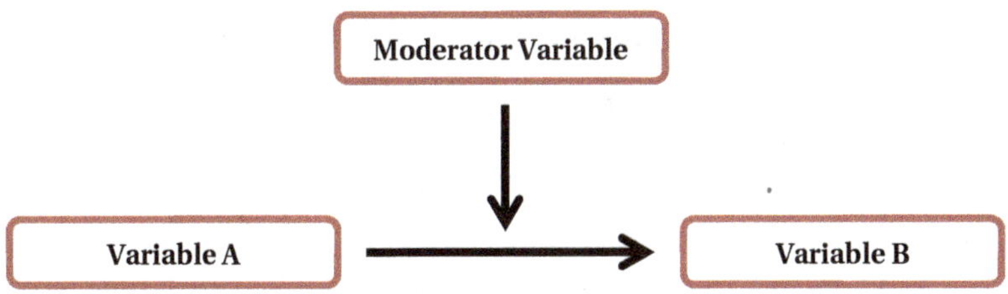

The second variable, the **mediator variable**, stands between the independent and dependent variables, and transmits the effect of the independent variable on the dependent variable (Creswell and Creswell, 2017). When the mediator variable is excluded, the relationship between the independent and dependent variables changes significantly. In other words, the independent variable (A) affects the value of the mediator variable, which in turn influences the value of the dependent variable (B) (Rosenstein, 2019; Weathington, Cunningham, and Pittenger, 2012). Figure 5.3 shows a research model with a mediator variable along with independent variable A and dependent variable B.

Figure 5.3. Relationship model: Example of a mediator variable

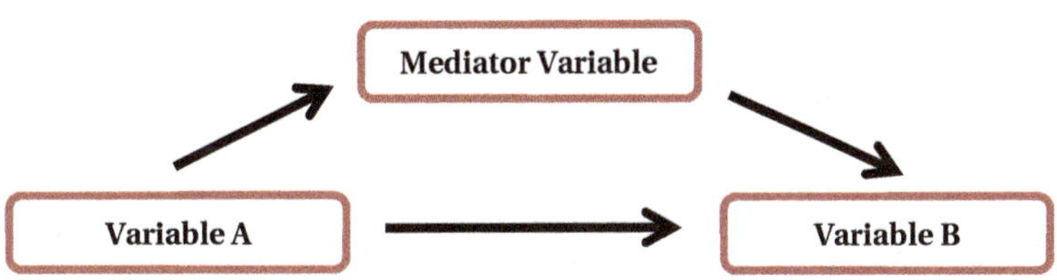

The mediator variable can play a partial or full role in explaining the relationship between the dependent and independent variables. In the case of a full mediator role, the effect of the independent variable (A) on the dependent variable (B) is only possible with the presence of the mediator variable. Therefore, variable (A) can only influence variable (B) through the moderator variable. This is mostly used in the social sciences to describe psychological processes such as emotions, thoughts, beliefs, attitudes, and behaviors, which are the dynamic features of people (Preacher and Hayes, 2008; Weathington, Cunningham, and Pittenger, 2012). The moderator variable effect can be tested using *moderated regression analysis* or *structural equation modeling* (Weathington, Cunningham, and Pittenger, 2012).

A research model might entail focusing on the network of relationships between variables. The concept of the variable and the types of variables should be understood well, since these may be important elements of a research model. Variables always affect, and are affected by, each other in a dynamic system such as a social, economic, natural, or cosmic system.

Variable Types According to Their Features

There are different kinds of variables according to their features: (1) *qualitative* properties and (2) *quantitative* properties. Further, they are specified by categories such as (1) *binary*, (2) *discrete*, and (3) *continuous* (Baur, 2011; Fraenkel, Wallen, and Hyun, 2011).

It is possible to distinguish variables as qualitative or quantitative regarding their descriptive properties. **Quantitative variables** refer to numerical features such as age, height, wage, and national income, while **qualitative variables** refer to aspects that cannot be described numerically, such as gender, marital status, quality, and brand preference. In other words, while quantitative variables express only a variation in quantity or degree, qualitative variables encompass variation in terms of type (Fraenkel, Wallen, and Hyun, 2011).

A variable's qualitative or quantitative characteristics also indicate the type of data involved. Color is qualitative. Therefore, the data we obtain about the color of an object is qualitative. For example, if we are talking about the types of houses we are researching (e.g., apartments, detached homes), we are dealing with a qualitative variable; if we are talking about their prices; we are looking at a quantitative variable. Qualitative (categorical) variables signal which category a unit will belong to due to its qualitative feature(s). Examples include the following:

→ Gender (female-male);

→ Types of cars (sports cars, sedans, SUVs);

→ Parcels of land (lakes, forests);

→ House models (diamond-ruby or circle-detached);

→ Tree species;

→ Tropical bird species;

- → Whether one invests in stocks;
- → Customer satisfaction (high-medium-low);
- → Type of institution (private-public, national-international, service-manufacturing);
- → Sector type (electronics, furniture, healthcare); and
- → Type of school (high school, technical school, online school).

When qualitative variables are evaluated statistically, they gain meaning by answering questions such as *"How many?"* or *"To what extent?"* For example, in a study where the entire population of Country A is considered the research population, questions that might be asked include, *"How many people are blue-eyed?"* or *"How much of the population is blue-eyed?"*

Quantitative variables are continuous or discrete. If there is a range for the data and the observation units can take any number (set of values) in a definition range (including decimals), we are dealing with a continuous variable (e.g., height, weight, speed, wages, the duration of machine repair). However, if the data can only be expressed as whole numbers, we are dealing with discrete data (e.g., age, exam grade, price, number of children, siblings, rooms, employees, accidents). In some exceptional cases, non-integer values can also be considered discrete data. For example, carbonated beverage producers generally bottle their products in the amounts of 0.222 liter (7.5 oz), 0.237 liter (8 oz), 0.355 liter (12 oz), 0.592 liter (20 oz), 1.25 liter, and 2 liters. Although these numbers are not all integers, discrete data are used (Bowerman, O'Connell, Murpree, and Orris, 2004; Dube, 2004; Freedman, Pisani, and Purves, 1980).

Traffic accidents are another example of discrete variables because the number of accidents cannot be expressed as 1.5. On the other hand, the height or weight of children in a class is a continuous variable because the observed units can take values such as 60.5 kg or 181.2 cm. In other words, the continuous variable is one that has many unlimited values (i.e., its value if infinite) that it can take freely within a certain range. Interest rates, electricity consumption, and precipitation are also examples of continuous data.

Note: Discrete and continuous variables are considered discrete data and continuous data in some sources. Both uses are correct. Age, educational background, or profession may be among the variables of a study. Each observation value for these variables represents a piece of data. Being discrete or continuous is a *feature* that characterizes the *data*. In this book, these features are accepted as describing variables since they characterize variables as discussed in some sources.

Types of variables by characteristic include the following (Baur, 2011):

- ⇨ A *binary variable* (indicates two categories, such as gender [male and female]) can be coded as 0 or 1.
- ⇨ A *discrete variable* (indicates more than two, but a finite number of categories, such as the number of children in a family or the number of continents on Earth [Africa, the Americas, Asia, Europe, and Australia]).
- ⇨ A *continuous variable* (indicates a large number of categories, such as income in $USD) can be coded through any value from 0 to an infinite number.

Data, Information, and Knowledge

Data consist of unprocessed, raw input and entail objective facts about events. Data are obtained through measurements (Turk, 2003), and can be qualitative or quantitative. Quantitative data can be gauged in definite units of measurement, and can be continuous or discrete (see variable types in Chapter 5). Qualitative data involve qualitative characteristics (Kundu, 2015).

Information is gathered and/or processed data. When the data are processed and classified, they become information, the purpose of which is to influence the receiver and to impact his/her thoughts and behaviors (Turk, 2003; Winograd and Flores, 1986). For example, the returns of investment instruments in 2018 include the following:

→ Treasury bond 4%	(Data 1)	
→ Gold 9%	(Data 2)	} Information
→ Stock market 6%	(Data 3)	

Knowledge entails the consequences of the facts that are reached through mental processes. Knowledge emerges due to the receiver interpreting information, and is formed in a person's mind (Turk, 2003; Winograd and Flores, 1986). In our example, the assessment that "*the investment instrument with the highest return is gold*" is **knowledge**, and is formed through human interpretation of **information**. The decision-maker decides where to invest, using the inference she/he has reached from this information.

There are many different types of knowledge and classifications of knowledge. Because there is no agreement on the classification of knowledge, I attempt to make a classification according to the domain of use as the following, similar to Ropohl (1997), Craig (2005), Pritchard (2014), and McCain (2019):

- ⇨ Everyday knowledge[2]
- ⇨ Technical knowledge[3]
- ⇨ Artistic knowledge[4]
- ⇨ Scientific knowledge[5] and
- ⇨ Philosophical knowledge[6]

This list can be extended, but this is sufficient here.

Other classifications of knowledge according to propositional or non-propositional information include the following (Cruz, 2006):

- ⇨ Propositional knowledge is based on a proposition. For instance, we might say, "Carlos knows that Oaxaca is in Mexico." There is a proposition in this case: "Oaxaca is in Mexico." It is simply known. Epistemology is mostly concerned with propositional knowledge.
- ⇨ Non-propositional (procedural) knowledge involves "how to" clauses. For instance, we might say, "Carlos knows how to ride a bicycle."

Types of data include the following (Kundu, 2015);

- ⇨ Primary data are not immediately available and are collected directly by the researcher.
- ⇨ Secondary data are immediately available and indirect.

On the other hand, there are some other types of knowledge, such as *practical knowledge* or *know-how* (knowledge of how to do something, how to perform some task), and *knowledge by acquaintance* or *experiential knowledge* (Stanley and Williamson, 2001).

Series are datasets where data are displayed according to their features. The most common type is a time series. Series are generally divided into two: time series and cross-sectional series. **Time series** are arranged to display observation

---

[2] Examples of information include the locations of bus stops, the departure times of buses, weather forecasts (by looking at the clouds), traffic, cooking, navigating social situations, and elevator usage.

[3] The information required to perform a specific operation, such as engineering, computer programming, and using a machine or equipment.

[4] This includes knowledge related to the creation, appreciation, and understanding of art forms such as painting, music, literature, and sculpture.

[5] This pertains to knowledge derived from the systematic study of the natural world, using empirical evidence, experimentation, and logical reasoning to formulate theories and models that explain phenomena.

[6] This involves the study of fundamental questions about existence, knowledge, values, reason, mind, and language, often employing critical analysis and rigorous argumentation.

values according to time (for multiple times) for a single observation unit/subject. They can be arranged on a daily basis, by year, every six months, every three months, monthly, or weekly. Other examples include a country's population growth by year, central budget by year, national electricity consumption by month or week, a company's sales revenue by month, season, or year, and the daily number of patients in a hospital (Kundu, 2015). **Cross-sectional series** are values taken for observation units at the same time (for a fixed period). In other words, all information is collected simultaneously (Everitt and Skrondal, 2002). For example, a cross-sectional series was used to represent the per capita national income of 2018 for member states of the Organization for Economic Co-operation and Development (OECD). Other examples include a company's sales revenue for the month of June by city, the average temperature in April by province, and countries' population growth rates in 2019.

Panel series (also known as panel data) are based on repeating several measurements in cross-sections in a given period. In other words, there are repeated observations on the same topic over the same time frame in panel data, such as longitudinal studies in which subjects are researched or tracked over time. The point to be considered here is that the data for a single year represent a single cross-section (Ajmani, 2009).

Data Collection Methods

Data collection is a process that consists of gathering data involving many variables (Rivera, 2020). The most basic ways of gathering data in scientific research are *observation, measurement,* and *experiments* (Jevons, 1958). Data collection methods[7] in scientific research include the following:

- ⇨ Surveys
- ⇨ Observation
- ⇨ Interviews
- ⇨ Document review
- ⇨ Social, psychological, laboratory, or thought experiments
- ⇨ Case studies

---

[7] Let us try to distinguish between *data collection methods* and *measurement tools.* Data collection methods, as a category, include general types of data collection techniques listed above, such as surveys and observations. Measurement tools refer to data collection tools specifically chosen in relation to the subject matter. For example, a questionnaire (subject-specific) is a measurement tool placed under a survey method. A survey is a measurement tool, just like a tape measure that measures our height.

A **survey** is a method of data collection based on a questionnaire, test, or interview (Mitchell and Jolley, 2009) that consists of many procedures. A survey is often used for descriptive research (Saris and Gallhofer, 2006), and is useful for answering the question, "*What do people plan to do?*" On the other hand, a survey might not always be useful for determining a cause-effect relationship. Survey-based studies mostly reflect *what*, but rarely reflect *why* (Mitchell and Jolley, 2009). For example, the distribution of responses on satisfaction with the economy, the government, or a given product can be measured via a survey (Saris and Gallhofer, 2006).

A successful survey must meet three main qualifications. First, a hypothesis should be identified so that the researcher knows what to measure. Second, the data collection tool (a questionnaire, test, or interview) should be selected correctly to ensure accurate measurement of people's thoughts, feelings, and behaviors. Third, the results should be generalizable to a certain group called a *population* (Mitchell and Jolley, 2009).

The digital age has provided researchers with numerous opportunities to gather data using multiple tools in a short period of time. These tools include survey platforms (e.g., Survey Monkey© or Qualtrics©), email, and digital voice and video recordings (Gay, Mills, and Airasian, 2011).

Observation is used widely and frequently in both the natural and social sciences; it is based on watching participants, especially to examine how natural, social, or psychological processes occur and develop in their usual flow. In other words, performing an observation helps one to understand how participants experience the natural environment without altering or manipulating it (Gay, Mills, and Airasian, 2011). Observation can be carried out through spectator or participant observation, depending on circumstances and availability (Patton, 2014).

A researcher can observe events (e.g., a worker performing a particular duty) and facts occurring around him/her by taking notes (open-ended) or using a checklist (closed-ended). Herschel referred to this as *passive observation*. However, when unusual combinations and factual conditions are artificially produced by interfering with environmental conditions, Herschel termed this *active observation* (Creswell and Creswell, 2017; Jevons, 1958).

> **"Two truths cannot contradict one another."**
> (Galileo Galilei n.d.)

An **interview** is another method for gathering data based on interactions between the researcher and the participant (Gay, Mills, and Airasian, 2011). There is very little difference between a questionnaire and an interview. In both cases, the researcher is interested in participants' responses to questions (Mitchell and Jolley, 2009). Interviews make it possible for researchers to collect important data that they could not otherwise obtain from observation alone (e.g., past events).

Interviews are more time-consuming and expensive than questionnaires. Whereas the researcher can administer a written questionnaire to many people at once, it is difficult to conduct an interview with more than one person at a time (Mitchell and Jolley, 2009).

Interviews provide additional interactions with the participants, and help the researcher to clarify questions that the respondents do not understand; the researcher can follow up with the participants regarding responses the researcher might not understand or did not expect. On the other hand, participants might give socially desirable responses that they think will make the interviewer like them or think well of them (Mitchell and Jolley, 2009).

Interviews can be conducted in two primary ways. They can be structured, which means they are based on a previously prepared set of questions. Structured interviews are formal and can be derived from data collected during past observations. Conversely, unstructured interviews are informal, but more than a regular conversation. Unstructured interviews do not entail answering predetermined questions, but rather allow researchers to inquire opportunistically about something that has emerged in discussion (Gay, Mills, and Airasian, 2011).

Researchers can record data in three ways for an interview: (1) They can take notes during the interview; (2) they can write down notes afterward; or (3) they can record the interview through audio or video (Gay, Mills, and Airasian, 2011).

Document review entails gathering materials such as *official reports* and *publications, archival files, journals, personal diaries* and *letters, photographs, maps, artistic works,* and *written responses* to open-ended surveys. Such data can be obtained from institutions, organizations, and clinical or official records (Gay, Mills, and Airasian, 2011; Patton, 2014).

Documents give researchers a chance to investigate the past. For this reason, document review is useful when they need to investigate prior events (Gay, Mills, and Airasian, 2011; Patton, 2014). However, researchers should pay attention to the accuracy and completeness of the documents provided (Patton, 2014).

In an **experiment**, the researcher controls the level of the independent variable(s) before gauging the level of the dependent variable(s) (Webster and Sell, 2007). In other words, the researcher manipulates at least one independent variable and controls other relevant variables, so that she/he can observe the effect on one or more dependent variables. The manipulation of an independent variable is the primary characteristic of an experiment, which is the most structured type of research of all data collection methods (Gay, Mills, and Airasian, 2011).

Experiments include *laboratory experiments* (in the natural and social sciences, as well as applied science), *social-psychological experiments* (in the social sciences; Mitchell and Jolley, 2009; Patton, 2014), and *thought experiments* (in the natural and social sciences).

In a social experiment, a group that receives a new treatment is called the *experimental group*, and a group that receives a different treatment (or is treated as usual) is called the *control group*. In social experiments, researchers should randomly select participants from a single, well-defined population and then randomly assign the participants to different treatment conditions (Gay, Mills, and Airasian, 2011).

An experiment can be designed to create a situation that is hard to observe in natural settings, to isolate one or several processes, to remove or minimize known confounding factors that normally cannot be separated from the phenomena of interest, or to create circumstances that do not yet exist or are known to exist in natural settings (Webster and Sell, 2007).

Social experiments involve examining social and/or psychological patterns to establish people's reactions to certain situations or events. In some cases, it may be difficult or impossible to apply social and psychological experiments for practical or ethical reasons.

Experiments can be concrete or intellectual. A **thought experiment** (*Gedankenexperiment* in German) can be carried out systematically around a fictional scenario or a group of variables (Gooding, 2005). Einstein frequently used thought experiments to better understand cosmic realities that were difficult to grasp.

Data errors endanger research findings. Potential errors can stem from *missing data, inaccurate data,* or *data collection methods* in quantitative research studies. Robust preparation is crucial before data collection to prevent or minimize data errors. For example, if data is collected by a questionnaire form, those who fill out the questionnaire may leave some parts blank in the form. It is very important for pollsters to notice this quickly and take action before it is too late. Alternatively, while coding the values, some researchers leave the data field blank for missing data or use an out-of-range value (Schneider and Deenan, 2004).

Related Terms and Concepts

The starting point of a scientific study is to ask a question or define a problem. A research study has *principles, criteria, rules, procedures,* and *structural features* (the research design) centered around a purpose. Let us explore the following terms and concepts related to scientific studies.

A **scientific premise** is a statement meant to support the conclusion of an argument (Cohen, 2011). In other words, a premise is a pre-assumption mostly based on the findings of prior studies that lead to the conclusion.

A result based on a premise allows us to reach a valid truth; this is what is meant by scientific proof. If our inference (the way we produce results) is correct, but our results are wrong, then at least one of our main premises is wrong. A premise might not always be true; it can also be false (Popper, 1972).

In terms of the social sciences, a scientific premise is that human activity can only be identified if we employ a set of intentional concepts (including *meaning, purpose, intention, significance,* and *norms*) since all human activities and cooperation are based on interactions and communication among individuals (Outhwaite and Turner, 2007).

A **scientific principle**[8] is the general compass that guides the researcher both in theoretical investigation and in the application of activities; scientific principles include *ethical principles* and the *principles of objectivity* and *impartiality*. Scientific principles are very diverse, and each field has its own particular ones (Marvin, 2011). Scientific principles and *scientific research principles* are different concepts. Scientific research is guided by six principles (McMillan, 2015: 7-8):

- ✓ To pose a significant scientific question that promises a benefit in line with one's needs, and that can be investigated empirically.
- ✓ To link research with relevant theory.
- ✓ To use method(s) that allow for direct investigation of the question.
- ✓ To follow a coherent, explicit, and evidence-based chain of reasoning.
- ✓ To replicate and generalize results across studies.
- ✓ To publish research to encourage professional reviews and critiques.

A theory in a particular scientific field can also have principles. For example, one of the key principles of Quantum Theory is *the uncertainty principle* (also known as the Heisenberg Uncertainty Principle), which states that it is impossible to know the exact position and momentum of a particle at the same time. This is because measuring one property, such as position, will inevitably disturb the other property, such as momentum. **Scientific rules** are mandatory regulations that guide applied sciences, such as math. They are developed specifically to produce consistent outcomes. For example, it is a rule to resolve

---

[8] The *Scientific Principles Dictionary* by Stephan Marvin (2011) provides comprehensive content on this subject.

the expression in parentheses first in mathematical operations. Let us consider the example below:

$$\frac{4}{10} - \left[\frac{5}{2} + \frac{1}{3}\right] \times \frac{2}{5}$$

If we recall our knowledge of math, the operation within parentheses is performed first. The second operation is multiplication (and/or division). The last operation is extraction. What we do here is follow a series of rules in mathematical operations. Different fields may have their own rules, which must be followed to produce consistent results, even when done by different operators.

Science tries to identify general laws that govern the world (Creswell and Creswell, 2017). **Scientific law** involves repetitive generalization, such as the *Law of Gravity*[9] or the *Law of Magnetism.* These are facts, not opinions, and are thus scientific laws. They are found not only in the natural sciences, but also in the social sciences and humanities, such as psychology (e.g., the *Laws of Memory*) and medicine (e.g., *general treatments* against particular diseases; Mitchell and Jolley, 2009).

Figure 5.4. Gravity is a natural (and thus a scientific) law

[9] Of course, the first question to come to mind is why evolution is called a theory, not a law. The answer to the question is that evolution proceeds with great complexity and is open to theoretical studies. It does not mean that evolution is contested; evolution itself is not a controversial area of study, it is the way of working of evolution that is studied and discussed.

Science is concerned with finding evidence and progressing through proof, which entails two possible conclusions in solving a hypothesis: (1) *verification* (confirmation) and (2) *falsification* (i.e., *disconfirmation*). Karl Popper (1902–1994), a well-known philosopher, pointed out the importance of a theory that is falsified in his book *Objective Knowledge* (1972). According to Popper, a theory that is not falsifiable is not a scientific one (DeWitt, 2010). The *Multiverse Theory* and *God's existence* are not falsifiable; hence, they cannot be theorized.

Scientific axioms are the basis of scientific propositions, which explain how the world operates (Gay, Mills, and Airasian, 2011). Axioms also present an approach that is *principle-like* or field-oriented. For example, an anthropological axiom asserts that one cannot understand another culture without understanding the language of the people who have that culture (Marvin, 2011; Patton, 2014).

Scientific postulates are *assumptions* of a particular field and are mainly based on theories and substantive knowledge in the field (Ragin, 2007).

An **assertion** is a scientific statement indicating a propositional generalization (Creswell and Creswell, 2017), and signals an assumption that is presumed to be true but is not yet verified (Gay, Mills, and Airasian, 2011).

A **hypothesis** is a specific *prediction (proposition) about the relationship* between two or more variables or *the parameter(s) of a population*; it is a prediction about a specific event in a relationship between variables. A hypothesis is based on a theory or prior research and needs to be verified or falsified (Creswell and Creswell, 2017; Weathington, Cunningham, and Pittenger, 2012).

The steps to be followed in hypothesis tests are stated below:

- ✓ Establishing a hypothesis (zero and alternative hypotheses);
- ✓ Determining the level of importance (α);
- ✓ Determining the size of the sample (n);
- ✓ Determining the test method (statistics);
- ✓ Determining rejection and acceptance regions in the sampling distribution (i.e., the critical table value according to the level of importance);
- ✓ Collecting data (on the population and the sample);
- ✓ Calculating test statistics; and
- ✓ The decision (accepting or rejecting the hypothesis).

In the realm of scientific inquiry, hypotheses manifest in various forms, encompassing the estimation of parameters, examination of correlations between variables, discernment of disparities among populations, and exploration of causal relationships. These delineated types of hypotheses serve as foundational constructs guiding research endeavors across diverse disciplines, facilitating the systematic investigation and elucidation of phenomena. Table 5.1 shows the features of the four general hypotheses used in empirical research studies.

Table 5.1. Features of the four general hypotheses used in empirical research[10]

Type of Hypothesis	Purpose of Research	Mathematical Expression of Hypothesis	Written Hypothesis
Estimate parameter	Use sample statistics to estimate population parameters	$M \approx \mu$	What is the average score of French language students taking a grammar test?
Correlation between variables	Measure two variables to determine their Correlation	$P > 0$ *or* $P < 0$	There is a positive correlation between time spent studying and exam grades.
Difference among populations	Compare two or more sample groups to determine if they are different	$\mu_1 > \mu_2$ *or* $\mu_1 < \mu_2$	Women, on average, live longer than men.
Cause and effect	Determine if the independent variable causes differences in the dependent variables	$IV \rightarrow DV$	High levels of sugar consumption cause diabetes.

It is not imperative to create a hypothesis in scientific research. In some cases, it is not even necessary. For example, formulating a hypothesis in descriptive surveys and ethnographic studies might not be necessary or sometimes inappropriate. In addition, creating a hypothesis can cause one to ignore other factors that may have an effect on the dependent variable (Fraenkel, Wallen, and Hyun, 2011).

A **theory** is an organized body of *concepts, generalizations,* and *principles* that can be subject to investigation (Gay, Mills, and Airasian, 2011). Theories provide researchers with an explanation or prediction about the relationship between variables. Thus, it is essential to understand the nature and use of variables in order to form questions and hypotheses. Theories have a place in quantitative, qualitative, and mixed-methods research (Creswell and Creswell, 2017).

Theories are knowledge networks developed to clarify the connections among variables, and are fundamental to science in terms of providing the

[10] Based on: Weathington, Cunningham, and Pittenger, 2012: 50

simplest explanation of a phenomenon. For example, cause-effect relationships between economic events are explained with the help of economic theories. Similarly, theories help us to grasp natural events. For example, the theory of evolution helps us to understand the nature of how living beings transform and adapt to the environment they live in. The general theory of relativity helps us to understand the curvature of space and time caused by objects in space. There is a close relationship between a theory and a hypothesis. However, they differ from each other. A hypothesis can be formed by starting from the available theory in the field.

Population and Sample

A **population** is a community of units with common features of interest to the researcher; it can consist of anything from all citizens of a country to all students of a math class in a school (Mitchell and Jolley, 2009). Populations can be any size and cover any geographic area. However, the entire group of interest (population) is rarely available for research (Gay, Mills, and Airasian, 2011). This is mostly due to factors such as limited sources and time. Hence, studies primarily focus on groups called **sample**, which represent the population well.

In statistical research, sampling is generally employed because it is not physically possible to include all units containing the entire population in the measurement (Kundu, 2015). Consequently, research is conducted on the sample, and the results are generalized to the population (Bickman and Rog, 2008; Gay, Mills, and Airasian, 2011). Using different surveys, collecting data based on the same sample size but over different sample units may produce different outcomes (Kundu, 2015).

In some cases, when access to the population is difficult or impossible, its definition can be narrowed. Thus, it is defined by the *accessible population*, which is thought to represent the *target population*. The sample is selected from the accessible population (Fraenkel, Wallen, and Hyun, 2011). For example, since it is not possible to reach the entire target market group (the target population) in Country A, only the target market groups (the accessible population) in a certain city or several cities can be defined as the population. The level of representing the target population of a chosen city or city group is very important. In order to achieve this, a city or group of cities (area sampling) that are considered to represent the target population the best should be selected.

A study's target population can be many different sizes. An entire country's population can be a research population (e.g., the voter population in election research). However, the population does not always have to be very general. Students from a region, city, university, or faculty, or students of only one department or who are registered in a particular course (e.g., students who attend a statistics class), can be the population of a study.

In a scientific study, all individuals (members) associated with the purpose of the study constitute the population. The researcher may desire an exact measurement (or full count, such as a census survey)—which requires reaching all the individuals (members) of the population to obtain the most accurate result—but this is not always possible owing to the need to reach too many units, as well as cost and time constraints. The purpose of *sample statistics* is to accurately estimate the parameter(s) of "*a numerical characteristic of a population or a model*" in the population. A group is then selected from the population as the sample, and the researcher tries to estimate the parameter for the entire population. The process of selecting some units (individuals) from the population is called **sampling**. In other words, the sample is a selected fraction of the population on the assumption that it will represent the population (Bickman and Rog, 2008; Everitt and Skrondal, 2002; Fraenkel, Wallen, and Hyun, 2011).

For successful statistical estimation, the findings of a measurement should represent the population in the best way possible, both quantitatively and qualitatively (Fraenkel, Wallen, and Hyun, 2011). The more appropriate the sample is in terms of representing the population, the more accurate the outcome will be. When the population is *homogeneous* (e.g., an animal population), the result derived from the sample represents the fact of the population more strongly. On the other hand, when the population exhibits *heterogeneous* features (e.g., humans), the result obtained from the sample represents the fact of the population more weakly. Thus, how the population is represented is important for an accurate statistical prediction. For this reason, it is crucial to select the sample that best represents the population. For example, population-related parameters (e.g., political trends) can be quite difficult to predict; many variables need to be considered. For instance, demographic variables such as age, gender, race, marital status, religious beliefs, education level, or socioeconomic and sociocultural variables (e.g., income group, settler status [rural, suburban, or urban]) should be considered while forming a sample. Thus, a sample should be created to reflect the features of the population in terms of distribution (Gay, Mills, and Airasian, 2011, Mitchell and Jolley, 2009).

Parameters and Statistics

Parameters refer to general data or values regarding a population. In other words, they are inferences about variables (reflecting a population's properties such as age, gender, level of education, income, motivation level, and membership status) that characterize the population. Parameters include values such as the *mean, standard deviation, variance,* and *ratio,* which are related to variables that characterize the population (Mitchell and Jolley, 2010; Rosenstein, 2019).

To better understand the differences between parameters and statistics, let us look at an example from the social sciences. In a study about students from Faculty A (the population), the mean age of the students in Faculty A is the parameter, while the mean age of the students from the Department of Business Administration is the **sample statistic**. In our study, Faculty A is the population, and if we know the students' mean age, we have the exact parametric value (the full measure). In contrast, if we do not know their mean age, then we can ask the students of a particular department to make a statistical estimation of the value of the parameter of Faculty A (as the population). The chosen department is the sample of our study, and the value of the sample is the statistical estimation of the parameter of the population, not the exact parametric value. If the sample is representative of the population, then we can infer that what is true of the sample is also true of the population (Mitchell and Jolley, 2010; Weathington, Cunningham, and Pittenger, 2012).

Statistical hypotheses are formed to describe population parameters (Weathington, Cunningham, and Pittenger, 2012). Although parametric values are commonly expressed quantitatively in statistical studies, this is not a requirement (Miller, Strang, and Miller, 2010; Mitchell and Jolley, 2010). Parametric values containing information about a population can also be qualitative, such as psychological, physiological, genetic (White and Lintzeris, 2010), structural, modal, or stylish. For example, if all the roses in a garden are red, *the color red,* which is a qualitative feature, is the parameter that represents the population.

In light of these explanations, a full-measure study (e.g., a census) gives a parametric value, not a statistical estimation. However, in widespread use, full-measurement values are also categorized as statistical data, which can be regarded as a broader use of **statistics**.

Experimental and Control Groups

The researcher tests the effects of the independent variable(s) on the **experimental group** (Creswell and Creswell, 2017), which is also known as the *treatment group.* In contrast, the **control group** is not subject to the independent variable test in order to eliminate the misleading effect of factors other than the tested variable on the test results (Dunning and Freedman, 2007).

The control group is also known as the *comparison group*, and in many instances, its members are instructed to maintain their daily (routine) activities. For this reason, the control group is not subject to the manipulation (or treatment) of the independent variable. Hence, experimental and control groups are only found in causal-comparative research studies (Creswell and Creswell, 2017; Gay, Mills, and Airasian, 2011).

The experimental and control groups may be different or not meet the needs of the research. For example, groups can be separated by age, ethnicity, intelligence, or income. Alternatively, groups can be similar to test a new treatment for a particular disease. The initial conditions of the dependent variable for each group should be considered carefully to ensure an accurate comparison such that only the treatment of the groups is different (Creswell and Creswell, 2017; Mitchell and Jolley, 2009).

For example, when we want to prove the hypothesis "*fatigue prevents learning*" through an experiment, *fatigue* will be the independent variable, and *learning* will be the dependent variable. We would investigate the relative effect of the independent variable (fatigue) on the dependent variable (learning).

In research, while the *control group* continues learning with *normal activities, the experimental group* is subject to a learning acquisition test *by increasing* and/or *decreasing fatigue*. The main reason for having two groups is that there is a need to produce an outcome of an ordinary situation, whereby the results from the experimental group can be compared. Thus, the measurement in the experimental group is compared to the measurement in the control group. Otherwise, it is not possible to interpret the findings obtained from the experimental group.

Not every study has an experimental group and a control group; they are only included in experimental research (Creswell and Creswell, 2017; Dunning and Freedman, 2007). Further, there are likely differences between the experimental and control groups other than differences related to the treatment. Such differences are called *confounding factors*. It is very important to be aware of them, as they can affect the results. Differences in the dependent (or response) variable (e.g., income, employment in social research, cardiac health in health research) may be due to treatment, confounding factors, or both (Dunning and Freedman, 2007)

Measurement and Scale

Measurement is the process of quantifying (or scoring) a specific property (aspect or characteristic) of a fact, event, or unit according to predetermined standards (Gay, Mills, and Airasian, 2011). For example, determining the distance between two cities is a measurement process, as well as a person stepping on a scale to see how much she/he weighs and calculating the area of a field.

Measurements made in scientific research will logically coincide with measurements we take in everyday life, which may sound a little more complex from a methodological angle. For example, in a study of social facts and events, a more abstract measurement instrument, such as a survey, could be used to compare the length or weight of a measurement unit. However, what is done is logically the same. When a researcher conducts a survey, she/he tries to gauge consumers' perceptions, employees' motivation levels, or citizens' expectations regarding the general economy.

Measurement is the most critical stage of a study. Using reliable instruments allows one to make objective observations (Weathington, Cunningham, and Pittenger, 2012).

> **"An experiment is a question which science poses to Nature, and a measurement is the recording of Nature's answer."**
>
> (Max Planck 1950)

The instrument should be able to determine the properties to be measured with the desired accuracy and without mixing these properties with other features. In other words, the instrument should be objective, and should be able to yield the same result in every repetition for the object/property it gauges (Lewis-Beck, Bryman, and Liao, 2004; Weathington, Cunningham, and Pittenger, 2012)

Measurement occurs after data are collected (Gay, Mills, and Airasian, 2011). A wide range of instruments is used in the social sciences. Some of these are implemented by the researcher, and others are employed by the participants of a study (Fraenkel, Wallen, and Hyun, 2011). Such instruments are exemplified in Table 5.2 below.

Table 5.2. Measurement instruments according to those who utilize them[11]

Completed by the researcher	Completed by the participant
Rating scale	Questionnaire
Interview schedules	Self-check list
Observation forms	Attitude scales
Tally sheet	Personality inventories
Flowchart	Achievement tests/ aptitude tests
Performance check list	Performance tests
Anecdotal records	Projective devices
Time-and-motion logs	Sociometric devices

[11] Source: Fraenkel, Wallen, and Hyun, 2011: 117

An instrument should be useful; when it is difficult to use, needs expertise, or has many requirements in labor, money, and time, it is not useful. In short, the instrument should make measurement easy and at minimal cost.

Types of Scales and Measurements

Stevens (1951) defined four types of scales: (1) nominal and (2) ordinal (categorical); and (3) interval and (4) ratio (quantitative). While nominal and ratio scales are objective (absolute), ordinal and interval scales are subjective. Among the measurement types, nominal (i.e., classification) and ordinal scales are qualitative. Interval and ratio scales are quantitative (Bowerman et al., 2004, De Gruijter and Kamp, 2007; Fraenkel, Wallen, and Hyun, 2011). I describe these four types of scales in detail below.

1. **Nominal scale**: Data are generated through naming, categorization, or classification. Nominal data are usually expressed in frequencies. Arithmetic operations cannot be performed. Only the number or percentage of members in a specific class can be determined. The gender of a survey's participants is an example of this type of scale. Only the equality (=) link can be checked on this scale. In other words, the gender of a participant is female (=) or not (≠). Equality can be sought in terms of certain characteristics (such as personality) possessed by individuals with the same car model. Suppose that there are 1,000 sales of sports cars per year in Country A. These cars will be purchased by customers with similar categorical features (e.g., gender, personality type).

   In addition, the following features are evaluated using nominal scales:

   → Gender (female/male)

   → Marital status (married/single/widowed)

   → Profession (teacher/doctor)

   → Title (professor/associate professor/assistant professor)

   → Place of residence (suburban/rural)

   → Ethnicity (white/black)

   → Faith (Christian, Muslim, Jewish, Buddhist)

   → Owns a personal computer (yes/no)

   → Type of car (sports car/sedan/SUV)

   → Movie broadcast ratings (G/PG/16)

   → Parcels of land (lake-forest)

- → House model (diamond-ruby)
- → Whether one invests in stocks (yes/no)
- → Type of institution (private/public, national/international, service/manufacturing)
- → Type of school (high school, technical school, online school)
- → Organizational strategies (cost leadership/differentiation)
- → Sector type (electronics, furniture, health care)
- → Alternative treatments for a disease (herbal/dietary supplements)

2. **Ordinal scale**: This is a qualitative scale in which categories are ordered. Ordinal variables can be numeric or not and reveal a level of importance or degree for the observed units. It is unclear how large the difference is between the measurements that are subject to the ranking. In this regard, they are subjective scales. Within them, a ranking by larger or smaller (</>) can be determined. For example, variables such as liking a brand (in a questionnaire based on low, medium, high enjoyment of a brand, or the ranking of brands according to the degree in which it is liked) represent ranking measurements.

   In addition, the following features are evaluated using ordinal scales:

   - → Order from smallest to largest
   - → Order from largest to smallest
   - → Length (from short to long)
   - → Level of education (primary to secondary to high school)
   - → Letter grades (A, A+, etc.)
   - → Title (assistant professor-associate professor)
   - → Income (low-medium-high)
   - → Customer satisfaction (low-medium-high)
   - → Quality (low-medium-high-perfect) and
   - → Participation in social activities (little-often-very often)

   Note: Qualitative variables are expressed as *nominal* and *ordinal*; the mean of measurements cannot be meaningfully calculated and cannot be used in mathematical operations.

3. **Interval scale**: This type of scale has equal distances between each number (or value) and the next number. Unlike nominal and ordinal

scales, interval scales can be expressed numerically and are hence suitable for mathematical operations. The starting point can be determined subjectively (Rosenstein, 2019). For example, 0 °C indicates a subjective starting point that does not mean there is no heat. In addition, interval scales can be transformed into each other. For example, a kilometer, a unit of length, can be converted to another unit of length, such as a mile, or a unit of temperature, Celsius, can be converted to another unit, such as Fahrenheit. A linear relationship can be established between equally spaced, defined units of measurement.

The following features are evaluated on an interval scale:

→ Level of education (if designed based on years: 1–5 years, 6–10 years, etc.)

→ Income level (if divided equally: €0–€1000, €1001–€2000)

→ IQ level (if equally spaced: 81–100, 101–120, etc.)

→ Temperature value (Celsius or Fahrenheit)

→ Length value (meters)

→ Length (if structured equally: 161–170 cm, 171–180 cm, etc.) and

→ Likert scale (such as 1–5 for satisfaction)

4. **Ratio scale**: The starting point is absolute and is usually taken as 0 (zero). Ratio scales are the most objective type of measurement and are divided into intervals from the absolute zero point. Hence, the proportional relationship between the measurements is objective.

The following features are evaluated using a ratio scale:

In the natural sciences:

→ Weight

→ Domain

→ Volume

→ Length

→ Height

→ Distance

→ Speed and

→ Temperature

In the social sciences:

→ Revenue (when expressed as an amount)

→ Course grade (1–100)

→ Number of children

→ Number of students/cars

Unit and Level of Analysis and Unit and Level of Measurement

The closely related concepts of *unit of analysis, level of analysis, level of measurement,* and *unit of measurement* are frequently used in research. It is important to distinguish their correct use.

The **unit of analysis** (e.g., the actor or unit to be studied, such as a manufacturing worker, costumer, student, human resources department [HRD], an organization as a case or practice, operation, behavior, event, idea, participant in a program, or a program) is employed to *analyze* the variables in a study and should be correctly identified as a part of the research design. Descriptions, findings, and conclusions are made for (the) appointed unit(s) of analysis (Creswell and Creswell, 2017; McMillian, 1999; Patton, 2014).

Level of analysis[12] refers to inferences (hypothesis testing, statistical inferences, or descriptive analysis) regarding the level of entity (the unit) or reality studied; for example, an individual, family, business unit, the department of a geographic region, society, or civilization. For example, in a study about income, the researcher should decide the level of analysis on which to base the study: individual (personal income) or household (household income). Researchers have used the individual, group, and/or organizational level of analysis in organizational research (Outhwaite and Turner, 2007; Patton, 2014; Schnake and Dumler, 2003). Both concepts are very closely related. The following table presents examples of the level of analysis. See Table 5.3 for the comparison of unit of analysis and level of analysis.

[12] In the social sciences, the distinction between the macro and micro levels is widely used. The micro level refers to lower units such as individuals, while the macro level refers to larger units such as countries, cities, and industries (Bryman, 2004).

Table 5.3. Unit of analysis and level of analysis

Unit(s) of analysis	**Level of analysis**
Individual(s)	Students of a faculty or patients suffering from specific illness
Group(s)	Groups in a work place
Math class(s)	All math classes in undergraduate programs of University A
Human Resources Department	HRDs of a particular industry
Company(s)	Companies running in a particular industry
Country(s)	One particular country or e.g. all Asian countries
Crisis	Financial crisis (e.g. 2008 global financial crisis)
Event(s)	Civil wars (e.g. Syrian Civil War)
Program(s)	The new educational program supported by AI

In case studies, cases are units of analysis that researchers aim to describe in depth (Patton, 2014; Platt, 2007). Units of analysis can be extended to field-oriented examples (Gay, Mills, and Airasian, 2011) as follows:

→ *In the social sciences and humanities*: individuals, groups, institutions, artifacts (objects, books, paintings, pictures), languages

→ *In political science*: political parties, local organizations of political parties, human rights associations, a district, city, geographic region, or an entire country

→ *In sociology*: social groups, social classes, local communities, women's movements, foundations and associations, institutions

→ *In psychology*: persons

→ *In social psychology*: groups, departments, or units of an organization

→ *In economics*: a business, an industry or sector, the national economy, national or international financial markets, multiple organizations

Level of measurement[13] similarly refers to the units that *measure* the variables. For example, let us assume we want to find out whether employees are satisfied with promotional opportunities at a company. We would administer a questionnaire item to the company's work units that asks, *"How*

[13] Statisticians use *level of measurement* to refer to types of scales. I am not a statistician, but I have attempted to distinguish between the concepts to prevent the reader from becoming confused when she/he comes across *level of measurement* in different sources. Thus, I prefer to use the concept as defined above, and to employ *types of scales* to replace *level of measurement* as it is used by statisticians.

satisfied are the members of your work unit with promotional opportunities?" This question *measures* job satisfaction *at the group level,* yet through individual perception. In contrast, an item that asks, *"How satisfied are you with promotional opportunities?" measures* job satisfaction *at the individual level.* As we see, for the same research task, researchers can direct the question to the individual or the group. For analysis, this can be treated as individual perceptions of a group-level phenomenon. Alternatively, this can be treated as an aggregate of the mean perceptions of a group or organizational-level phenomenon. According to our questions (directed to the heads of units or all members of work units), our level of analysis is at the organizational level (even if the level of measurement is different or not at the individual or group level, for example), because we want to see whether the company's employees are satisfied with promotional opportunities. Thus, researchers can decide to extend the level of analysis from the group level to the organizational level using individual-level data (Schnake and Dumler, 2003).

The **unit of measurement** refers to how a variable is measured. For example, when using participants' age as a variable, the unit of measurement is commonly the *year.* The unit for income is *currency* -e.g., the dollar- (Bryman, 2004).

Paradigms and Modeling

According to Kuhn, a **scientific paradigm** is the core assumptions shared by a group of scientists in a research domain. A paradigm is not the same as a theory. Paradigms are larger than theories; thus, multiple theories can operate within a paradigm. Paradigms contain *schemas*[14] and *mental models*[15] that can direct the researcher in how to conduct a scientific study. Scientists studying one paradigm are somewhat immune to the arguments made by scientists studying a different paradigm. Consequently, even though paradigms are a useful compass for researchers, sometimes a critical attitude toward the dominant theory or paradigm can be useful for a scientific study (Ewoldsen, 2017). Paradigms indicate a worldview as denoting a basic set of beliefs that guide action. Because of this, all fields of science have philosophical paradigms related to each respective field. Paradigms are also related to the philosophy of science as a research method or are field-oriented. For example, the American Psychological Association (APA) style is a social science paradigm (Creswell and Creswell, 2017; Outhwaite and Turner, 2007).

---

[14] A schema is a mental representation that stores knowledge as interrelated pieces of information (Ewoldsen, 2017).

[15] Mental models are used to process, organize, and comprehend incoming information (Ewoldsen, 2017).

Scientific models are artificial designs (representatives) that illustrate a process, structure, network, event, or causal relations between variables. Scientific models can also perform an *explanatory* function through illustration in some cases. The Tree of Life is an example of a scientific model in biology. In another instance, social networks are a model of social life. Business models, mathematical models, architectural models, or research models are examples in the applied and basic sciences (Bailer-Jones, 2003; Bokulich, 2011).

Scientific models are about phenomena and are closely related to propositions (Bailer-Jones, 2003). The researcher should relate propositions or research questions to the research model in the study.

CONCLUDING REMARKS

Science is an endeavor related to many concepts and topics, and it is not easy to master all of them. That's why you can sometimes see people working in science get stuck on a single word or a single stage of a process. This is difficult to make sense of from the outside, but those involved in scientific research studies understand why. Sometimes, it may be difficult for the researcher to find the answer to the concept or process she/he is stuck on. For this reason, using more than one source is often a useful strategy to both find the answer you are looking for and specialize in the concept or process.

SUMMARY OF CHAPTER 5

A variable is an indicator of a state, feature, or element of a population. It is very crucial to comprehend the concept of the variable, to execute a research study properly. Two types of variables are distinguished; according to their features, and according to their roles in the research studies. Especially in the process of modeling, variables are the main elements of the research and should be properly defined according to both their feature and role in the research study.

Data collection methods are processes related to gathering data about variables in the research study. A researcher should choose the right data collection method in according to the requirements of the research. There are many ways to collect data, such as, *survey, observation, interview, document review, experiments* (e.g., social, psychological, laboratory), or *case study.*

A population is a community of units with common features. A population is a concept that is defined in line with the purpose and needs of the research study. Because of this, a researcher can narrow or expand the definition of a particular population in accordance with the purpose and needs of the research study. On the other hand, a sample refers to a part of the population that is expected to represent the population with maximum accuracy. Sampling is the process to determine the best (the most representative) selection of parts of a population to be included in the scientific study.

Parameters indicate general data or values regarding a population. In other words, parametric values represent the population in terms of certain features. On the other hand, statistics are the estimations or inferences regarding the population derived from samples that are anticipated to represent the population best. A researcher focuses on the sample to make a statistical estimation or inference when she/he knows that the whole population is not reachable for the measurement.

There is a strong bond between paradigms and modeling in the process of pursuing science. A scientific paradigm indicates assumptions shared by a group of scientists in a research domain and contains schemas and mental models to represent their shared assumptions. Researchers use those existing schemas and mental models derived from the assumptions in the paradigm, to design their new research model. Modeling is the design process of a research study. Scientific models are artificial designs and show a process, structure, network, event, and causal relations between variables, etc.

Chapter 6

SCIENTIFIC RESEARCH METHODS

Before starting to read the chapter

There may be countless ways of conducting a research study. But probably, some of them are much more capable of catching the truth. Researchers strive to find the best way of collecting and processing data in accordance with the field and subject of study. This does not mean that there is a single method best suited to a particular field or subject of study in social sciences. On the contrary, method may vary depending on the purpose of the research, its limitations, and the size of the accessible sample. Therefore, it is common practice to use different research methods on a particular subject, as is often done in the field of business administration. For this reason, researchers are advised to be familiar with more than one method. Because sometimes they may encounter research topics and conditions that require the use of more than one method at the same time.

The purpose of this chapter is to introduce readers to the main lines of research methods commonly used in social sciences. The content here presents the general outline of the methods. Each method also requires further reading from different sources. Besides, while some research methodology books focus on the applications of a particular type of method, some books present research methods for a particular field of science and provide examples.

This chapter presents scientific research methods. The chapter contains four subsections. The first three respectively introduce *quantitative methods, qualitative methods,* and *mixed methods.* The last subsection of the chapter considers *other classifications of research methods* according to the outcome features of the research study.

SCIENTIFIC RESEARCH METHODS

First, it is necessary to make a distinction. Scientific research methods and analysis techniques are related but separate topics. A *research method* begins at the stage of determining the purpose involved in the research process; it refers to a general practice that holds each stage of a study to certain standards from the beginning of the research, including the identification of keywords, assumptions, scope, and limitations. However, the *analysis method* entails how to analyze the data obtained.

Scientific studies are subject to differentiation based on the *characteristics of the chosen research method* and the study's *outcome feature, purpose,* and *intervention* or *participatory features.* Many events, structures, and situations are intricate in natural, human, or social events. Many factors (and reasons) produce these intricate events, structures, and situations; it is not easy to identify and examine all of them at once.

Researchers must be aware that every method or methodological approach has both strengths and limitations (McKnight, 2011). The type of research that takes place in a certain classification can be included in another classification due to its features. I present them in the list below. In the upcoming sections, I cover these concepts under related titles. Scientific research methods are divided into three types according to the *operational features* as follows. I explain them in the following pages in detail:

1. **Qualitative research**
2. **Quantitative research**
3. **Mixed methods**

While qualitative research seeks to answer "**why**" and "**how**" questions, quantitative research seeks to answer questions such as "**how much**" and "**how often.**" Qualitative research develops concepts and theories at the end of a study, quantitative research begins with theories and experiments. Qualitative studies are inductive; that is, they reach the result/reality based on a theory (Fraenkel, Wallen, and Hyun, 2011). Qualitative methods are mostly descriptive, while quantitative ones are explanatory (Patton, 2014). On the other hand, there is no consensus on classification of scientific methodology even present day. Several opinions are reflected in dedicated methodology books.

Quantitative Research

Quantitative research can be defined as research based on numbers. The most distinctive difference of quantitative research is the interpretation of statistical inferences. Quantitative methods use numeric scales for measurement. Lemon et al. (2010) put forward a simple example to exhibit the difference between quantitative and qualitative research. Quantitative research is very similar to a human's use of a meter to measure her/his height. The distinction between quantitative and qualitative research arises from this point. Qualitative research seeks people's feelings about their height. Remember that you can measure a human's height by meter or just asking to her/him what her/his height is. In the first one, the researcher conducts the measurement, and in the second one participant answers the question researcher. Both ways are appropriate.

Quantitative research aims to expose the statistical descriptions of the variables studied. In quantitative research, part of the population is taken as the sample and an estimation is made about the population under a certain error margin (Kaplan, 2016). More specifically, quantitative research aims to describe situations, events, and concepts (e.g., social networks) by *descriptive research*; examine relationships among variables (e.g., political crisis and economic growth) by *correlational research*; determine the effectiveness of a treatment, intervention (e.g., a new medical treatment) by *experimental research*, or problems/effectiveness of a process/activity (e.g., electronic voting system) by *observation*. Quantitative research can be designed as experimental, quasi-experimental, and non-experimental *–which indicates descriptive research, causal–comparative, and correlational research-* (Castellan, 2010; Grove and Gray, 2018; Rutberg and Bouikidis, 2018; Weathington, Cunningham, and Pittenger, 2012).

A quantitative study can be experimental or non-experimental. In a non-experimental study, the researcher is unable to manipulate or control any factors through intervention or treatment that may influence the participant's (subject's) behavior or performance, since it has already occurred or cannot be influenced (McMillan, 2015; Rutberg and Bouikidis, 2018). Descriptive research is suitable when little knowledge is available, and other research types can be employed as knowledge increases (Grove and Gray, 2018).

Qualitative data used for quantitative research can also be grouped with an appropriate classification (or categorization), and frequency distributions can be shown. Examples of frequency distributions include the number of letter grades (categorically ready, since the numerical grade ranges that correspond to the letter grade are predetermined) that students receive in a math class, the average duration of education of a group (intervals can be determined subjectively by the researcher if they will be categorized by year), or the number of broadcasts on national TV channels according to the type of broadcast (such as series, documentaries, news, and discussions).

General problems associated with quantitative research are *internal validity, external validity, and reliability*, which are discussed in depth in Chapter 7 (Buhringer and Sassen, 2010). Additionally, a researcher should *describe the dataset* and *categories* studied, which can accurately be done via statistical programs. *Randomization* and *control of extraneous variables*, and *researcher's bias* are other factors that may cause potential problems. Researchers should also avoid *poor research design* (Lemon et al., 2010; Fraenkel, Wallen, and Hyun, 2011).

Types of Quantitative Research

Among quantitative studies, surveys, as well as correlational and causal comparative research, are descriptive (and also predictive). In contrast, experimental and quasi-experimental studies are interventionist and practical. The most prominent other features of quantitative research include the following (McCusker and Gunaydin, 2015):

- ⇨ Researcher uses tools, such as questionnaires or equipment, to collect numerical data
- ⇨ Data is in the form of numbers and statistics
- ⇨ Analysis results and inferences methods are conveyed statistically
- ⇨ More standardized methods for processing data

The distinction of quantitative research methods presented here is organized according to their features presented above.

1. Surveys

Survey research is a method to test hypotheses on a particular subject. Surveys include collecting data by asking people for their opinions about subjects such as their history, habits, knowledge, attitudes, behavior, etc., in order to estimate the characteristics/parameters of the studied population (Everitt and Skrondal, 2002). A survey is conducted on a sample that represents the population to determine its characteristics with tools such as interviews and questionnaires (Fraenkel, Wallen, and Hyun, 2011; Gay, Mills, and Airasian, 2011). Because of this, it is usually considered as a sample survey.

2. Correlation Research

Correlation research is a quantitative method based on examining the relationships between variables in a single group; it often reveals cause-effect relationships. Correlational studies may aim to present explanatory findings or are used to make predictions. Techniques such as *simple linear regression, multiple regression,* the *multiple correlation coefficient* (*R*), *factor analysis,* and *path analysis* are examples of methods used in correlational research. Correlational studies require at least two quantitative variables, as well as a score for each variable for each subject.

3. Causal Comparative Research

Causal comparative research is based on comparing groups (or individuals within a group) with different experiences to identify the causes or consequences of group membership. While doing this, the researcher looks for a relationship between variables to determine the causes or consequences of the differences. However, causal-comparative studies are not as effective as experimental research in establishing causes. Unlike experimental research, the independent variable is not manipulated. The measurement is made as usual regarding the independent variable. The differences have already appeared between (or within) groups. Further, causes and consequences emerge. Therefore, causal-comparative research is also referred to as *ex post facto* (from the Latin for "after the fact") research. Causal comparative studies require at least one categorical (qualitative) variable (such as group membership) (Fraenkel, Wallen, and Hyun, 2011).

The most distinct difference between causal-comparative and correlational research is that causal-comparative research involves two (or more) groups of participants with one grouping variable, whereas correlational research involves two (or more) variables and one group of participants. In other words, causal-comparative research aims to determine differences among groups under a causal effect, whereas correlational research aims to determine whether a relationship exists between two or more variables (Gay, Mills, and Airasian, 2011).

4. Experimental Research and Observation

Experimental research is based on manipulating the independent variable (Fraenkel, Wallen, and Hyun, 2011). In experimental research, participants are randomly assigned to different groups (McMillan, 2015). Researchers manipulate one or more independent variables in experimental studies to examine the individual and collective effects on the dependent variable. Experimental research can be used in both the social and natural sciences. In the social sciences, experimental studies are used to understand human behavior (Rogelberg and Olien, 2014).

Experimental research can also be conducted *quasi-experimentally*, in which participants cannot be assigned randomly. In that case, the assignment is based on self-selection, happenstance, or convenience (McMillan, 2015; Rogelberg and Olien, 2014). A research study can be conveyed experimentally or via observation through the following designs (Rogelberg and Olien, 2014):

⇨ Laboratory-based experimental research

⇨ Laboratory-based quasi-experimental research

⇨ Laboratory-based passive observation

⇨ Field-based experimental research

⇨ Field-based quasi-experimental research and

⇨ Field-based passive observation

The purpose and limitations of the research should be considered in deciding on the study design (Fraenkel, Wallen, and Hyun, 2011; Rogelberg and Olien, 2014).

Qualitative Research

Qualitative research does not involve numerical values, but rather words and their meanings. There are three types of commonly used data collection methods in qualitative research: interviews (including focus groups), observation, reviews of written documents, field notes, photographs, audio recordings, videotapes, diaries, personal comments, memos, official records, textbook passages, etc. In other words, qualitative research is conducted by extracting words from texts (documents, transcripts) or notes taken during interviews, and observations, or by taking pictures and using other visual materials (Bogdan and Biklen, 1997; Fraenkel, Wallen, and Hyun, 2011; Goetz and LeCompte, 1984; Kaplan, 2016; Patton, 1987; Yildirim and Simsek, 2011).

It is possible to divide interview methods into subgroups. These include open-ended vs. structured according to the characteristics of the instrument used; individual vs. focus groups according to the number of people interviewed; or participant vs. non-participant according to the researcher's location (Yildirim and Simsek, 2011).

Qualitative research aims to describe the unit of analysis in depth, holistically, and in context (Gay, Mills, and Airasian, 2011; Patton, 2014). Thus, qualitative research may include a detailed description of a place or situation, or the actions of people, such as gestures, jokes, and conversational gambits, depending on the needs of the research study (Chigbu, 2019; Fraenkel, Wallen, and Hyun, 2011). In qualitative research, since the data are qualitative, they are categorized (classified) generally using frequency distributions (Creswell and Creswell, 2017).

Qualitative methods have always existed in the history of science. But one of the prominent works on discussing qualitative methods had been done by Bronisław Malinowski in 1922 in which there was no direct address to the meaning of "qualitative." Even today there is no consensus on what qualitative research is. But many philosophers focus on the notion of "quality." Quality is about what something is or has (Aspers and Corte, 2019).

Qualitative research is largely associated with two philosophical approaches: (1) phenomenology and (2) hermeneutics. But sometimes philosophical basis is ambiguous and the methodological basis of qualitative research can be founded on empirical studies as in Grounded Theory (Dahlberg, 1995).

Some quantitative studies consider qualitative research only as a way of generating hypotheses, and then testing them using quantitative analysis. In other words, in this case, qualitative research is only a supportive tool for quantitative research (Kaplan, 2016).

Qualitative research is not free from the high standards of scientific inquiry regarding validity, reliability, and objectivity. Qualitative methods, especially in the social sciences, may display a tendency to engage with subjectivity. The problems of the validity and subjectivity of qualitative research stem from the varied, unstandardized practices and the complex nature of qualitative studies (Holliday, 2012). *The generalizability of cases, the number of cases needed,* and *standardization* are the general problems associated with qualitative research (Aspers and Corte, 2019). Maxwell (1992) described five guiding standards to validate qualitative research as follows:

- ✓ Descriptive validity indicates factual accuracy
- ✓ Interpretive validity is related to researcher's concern about the participants' perceptive
- ✓ Theoretical validity refers to the ability of scientific research to explain the phenomenon studied generalizability is related to what extent the findings of scientific research can be generalized within the subject. *Internal generalizability* is more critical than *external generalization* for a research study due to relating the findings to the studied population
- ✓ Evaluative validity refers to presenting data

Types of Qualitative Research

Qualitative research has some distinctive features. The first feature distinguishing qualitative research from quantitative research is the feature of data. In qualitative research data used is qualitative. The most prominent other features of qualitative research include the following (Bogdan and Biklen, 1997; Fraenkel, Wallen, and Hyun, 2011; Glaser and Strauss, 1967; Goetz and LeCompte, 1984; Patton, 1987):

- ⇨ Sensitivity to the natural environment
- ⇨ The researcher's participant role
- ⇨ A holistic approach
- ⇨ Revealing perceptions
- ⇨ Flexibility in the research design and
- ⇨ Inductive analysis[1]

---

[1] Unlike the positivist approach—which examines cause-effect relationships depending on a particular institutional structure or deductive principle—the principle of induction

The distinction of qualitative research methods presented here is organized according to their features presented above.

1. Ethnographic Research

Ethnographic studies help the researcher to describe and interpret cultural or social groups. Ethnographic researchers use observation, interviews, and other analyses to understand the nature of a cultural or social group, and spend time documenting or depicting their daily experiences by observing or interviewing individuals and other related people (McMillan, 2015). Data can be collected through interviews, observation, and/or active participation in ethnographic research. Data analysis is completed by describing and interpreting data and theme development (Chigbu, 2019).

> "Ethnography is a design of inquiry coming from anthropology and sociology in which the researcher studies the shared patterns of behaviors, language, and actions of an intact cultural group in a natural setting over a prolonged period of time. Data collection often involves observations and interviews" (Creswell and Creswell, 2017: 62).

Ethnographic research has been developed by anthropologists who wish to study a society or a part of it. The word "ethnography," from *ethnos* (nation) and *graphy* (writing in terms of a description), involves directly watching a group's behavior and describing the group based on one's observation. Ethnographic research aims to establish a direct relationship with group members, explain the group's cultural structures, and describe the behaviors and experiences that make up these structures (Outhwaite and Turner, 2007).

2. Historical Research

Historical research involves the past; it is based on collecting data by *examining the documents* of a certain period or by *interviewing people* who lived at that time. Historical research seeks to answer the question, *"What happened in the past?"* (Fraenkel, Wallen, and Hyun, 2011).

Data collection can be done using interviews, surveys, and/or documents (text records, archival research) in historical studies. Data analysis is mostly completed by describing how events developed (Anderson, 2007; Chigbu,

---

is dominant in qualitative studies. The researcher tries to reveal the central themes related to the problem she/he has analyzed based on the descriptive and detailed data she/he has collected, in order to give the data she/he has gathered a meaningful structure; that is, to create a (grounded) theory rooted in these data. Even though these central themes are sometimes predetermined, there is flexibility in the qualitative research process whereby these themes are always open to change.

2019). On the other hand, repeated phenomena such as census schedules, vote tallies, vital records (e.g., birth, death, marriage), crime reports, and judicial and fiscal records, enable computer-based data entrance for making analysis and inferences by using quantitative methods (Anderson, 2007).

3. Action Research

Practitioners have problems in many areas, such as education, the business world, health care, and teamwork. Problems can emerge in performance, leadership, or conflict. Action research is usually conducted due to the existence of a problem that one aims to solve or improve. In action research, researchers implement interventions to develop better practices regarding the problem. Therefore, action research starts with diagnosing a problem (Tomal, 2010).

Action research is a form of applied research/science, and its targets change; it is carried out to solve problems and enhance processes (Fraenkel, Wallen, and Hyun, 2011; Tomal, 2010; Winter and Munn-Giddings, 2001).

Action research involves practitioners who work alone or with a researcher to understand and solve problems arising in practice; it involves an action plan. In this regard, it is a type of intrusive work. For example, in the school environment, action research helps teachers, administrators, and other specialists who are responsible for children's education to better understand the educational processes and practices they have created, as well as to improve processes, practices, and results. In this way, research and implementation are intertwined. Research outcomes can be implemented immediately, and new findings can be obtained by directly investigating the results of the application (Fraenkel, Wallen, and Hyun, 2011; Glesne, 2011). Action research can be based on qualitative methods (e.g., interviews, observations, documents) and can also use quantitative data (Tomal, 2010; Winter and Munn-Giddings, 2001).

4. Phenomenological Research

Phenomenological research entails understanding the essence of a phenomenon (McMillan, 2015); it focuses on describing a phenomenon or concept that has been experienced by several individuals, such as insomnia or anger. Researchers collect data from people who have experienced the phenomenon and develop a composite description of the essence of the experience for all the individuals (Creswell, 2007). Data can be collected through interviews, surveys, and/or observation. Data analysis is completed by describing experiences, examining meanings, and developing themes (Chigbu, 2019).

5. Grounded Theory Research

Grounded theory is a qualitative research technique that progresses through continuous interactions between data and emerging meanings. Based on raw data, it aims to reveal a theory through an inductive approach (Fraenkel, Wallen, and Hyun, 2011). Data in grounded theory research can be collected through interviews and questionnaires, and data analysis is completed by coding data, categorizing themes, and describing implications (Chigbu, 2019).

Grounded theory purports to develop theories via induction based on data. Researchers devised this name because they tried to extract theories from embedded data, so there is no developed theory before a study begins. After the researcher gathers data, a generalization is made. In this respect, grounded theory emphasizes continuous interactions between raw data and the comments that the researcher extracts from the data. Hence, researchers use the constant comparative method, which is based on the relationship between the data collected and interpreted, and the theory that is developed. Potential categories are created to group the items (created), and the work continues until the data and theory fully overlap. It is an uninterrupted process that continues between the data and theory. Data are collected and analyzed; a theory is proposed; more data are collected; the theory is revised; more data are gathered; the theory is further developed, clarified, revised; and the process continues like this (Fraenkel, Wallen, and Hyun, 2011).

Grounded theory emerged from the work of Glaser and Strauss (1967) in the health sciences. Glaser and Strauss examined the interactions between healthcare professionals with terminally ill patients. Based on the data they obtained, they identified various dimensions and results and developed new concepts and explanations regarding the interactions between healthcare workers and patients. They put forward an explanatory theory based on their data. They explained the emotional phases of terminally ill patients as "denial," "anger," "acceptance," and "reconciliation." These stages have been observed in most patients, and have thus become a theory of this process. This type of theory can constitute an important conceptual basis for healthcare professionals. For example, with this theory, healthcare professionals can attain an understanding of the problems and abnormal behaviors of their terminally ill patients and direct their interactions with these patients accordingly (Hancock, 2004).

6. Case Studies

Case studies involve qualitative data in which researchers focus on a unit of study in detail, known as a bounded unit or system -e.g., individual workers, the units of an institution, the department of a company, a company in business life, an individual teacher, a classroom, or a school- (Mitchell and Jolley, 2010; Gay, Mills, and Airasian, 2011).

In a case study, the most important initial step is to identify the right case to answer the research question (Patton, 2014). In a case study, questions are adapted for the particular case selected. The generalization of the study requires theory-based arguments (Maxwell, 2013). "*Case studies are often talked about as a product. The case study stands on its own as a detailed and rich story about a person, organization, event, campaign, or program*" (Patton, 2014: 394).

Case studies can be used for different units of study such as individuals, departments, or organizations. Case studies are frequently used, especially in medicine and clinical psychology, as well as sociology and anthropology (Platt, 2007).

Single or multiple case studies are focused on fully understanding cases in order to place them in a broader context. In this sense, case studies are *heuristic.* Multiple case studies provide researchers with the opportunity to compare cases (Gay, Mills, and Airasian, 2011; Platt, 2007).

> "Case studies are a design of inquiry found in many fields, especially evaluation, in which the researcher develops an in-depth analysis of a case, often a program, event, activity, process, or one or more individuals. Cases are bounded by time and activity, and researchers collect detailed information using a variety of data collection procedures over a sustained period of time" (Creswell and Creswell, 2017: 62).

In a case study, multiple qualitative data collection methods can be employed, such as observation, interviews, and document review (Gay, Mills, and Airasian, 2011). Case studies can involve individuals, institutions, groups, or environments, and can apply a quantitative or qualitative approach. Both approaches aim to produce results related to a given situation. The most basic feature of qualitative case studies is that they investigate one or more cases in depth. In other words, factors related to a situation (e.g., environment, individuals, events, processes) are investigated with a holistic approach, and the focus is on how they influence the situation and how they are affected (Yildirim and Simsek, 2011). Data analysis is completed through the detailed identification of themes and the development of narratives in case studies (Chigbu, 2019).

In case studies, more than one data collection method is generally used; the techniques include interviews, observation, documents, and physical inspections. In this way, the researcher tries to obtain a rich and broad array of data. Since situations differ from each other, the results are not generalized. However, the findings obtained for a situation are expected to provide examples and experiences to understand similar circumstances (Chigbu, 2019; Yildirim and Simsek, 2011). Thus, criticism is often directed toward case studies, with some scholars pointing out their limitations in terms of generalizing findings (Platt, 2007).

7. Narrative Studies

Narrative studies examine people's experiences regarding a topic or situation through stories they have experienced. Examples include autobiographies, biographies, life stories, and personal story studies (Fraenkel, Wallen, and Hyun, 2011). Data are collected through interviews and documents; data analysis can be completed through storytelling, content review, and theme development (Chigbu, 2019).

8. Critical Research

Societies have experienced historical problems of domination, alienation, and social struggles. Critical research focuses on the kinds of experiences facing societies and social institutions. Critical research critiques the subject-society and aims to encourage society to overcome constraints and envision new possibilities (Creswell, 2007).

Critical research is carried out in the social sciences and goes far beyond fault-finding. The core of critical research is critical thinking. Critiques center on examining social institutions, ideologies, discourses, and forms of consciousness in terms of representation and domination. Critical research also helps one to understand social realities that follow cultural and linguistic conventions. Critical research is based on challenging (rather than confirming) that which is established, and disrupting (rather than reproducing) cultural traditions and conventions. Hence, critical research mostly focuses on the role of mass culture and communication in social reproduction and domination (Alvesson and Deetz, 2020; Outhwaite and Turner, 2007).

Mixed Methods Research

Mixed methods research uses design characteristics from both qualitative and quantitative approaches (i.e., it combines qualitative and quantitative data) in a single study. Mixed methods research designs allow researchers to avoid or minimize the weaknesses of qualitative and quantitative methods when used alone (Gay, Mills, and Airasian, 2011; McMillan, 2015).

In a mixed methods study, qualitative data tend to be open-ended, without predetermined responses, while quantitative data usually include close-ended responses (Creswell and Creswell, 2017). Given the blended nature of a mixed methods study, considerable time and resources are required to carry it out (Creswell and Creswell, 2017; Gay, Mills, and Airasian, 2011).

Other Classifications

Another distinction divides scientific research into two branches according to the *outcome feature* of the research. These include the following (Fraenkel, Wallen, and Hyun, 2011):

i. **Basic research** contributes to the scientific background. Basic research aims to produce new knowledge and develop theories about previously conducted research. It involves clarifying the processes underlying the phenomenon or event observed. In basic research, the hypothesis is often expressed as a theory.

ii. **Applied research** is conducted to examine the effectiveness of an application. Applied research studies are based on the investigation of the usefulness (through intervention) of the theories put forward in practice.

As readers can easily realize the categorization presented above is similar to the content discussed in Chapter 1 under the title –Basic Sciences and Applied Sciences-. The emphasis here is on the type of research. Recognizing the types of research helps us determine how to design the research based on its purpose.

There are also general categories in which research methods are included. According to the *outcome feature* of research, these categories include the following (Fraenkel, Wallen, and Hyun, 2011; McMillan, 2015):[2]

1. **Descriptive studies** describe a state (state of affairs, situation, or event) as thoroughly and carefully as possible. One of the best examples of descriptive research is found in botany and zoology, where each plant and animal species is meticulously defined, and knowledge is organized into useful taxonomic categories. Taxonomic and typological studies are descriptive studies.

2. **Associational research** investigates relationships; for example, correlational and causal-comparative research. Correlational studies examine relationships among two or more variables (e.g., the correlation

[2] Statistical operations encompass two primary domains: descriptive statistics, which involve summarizing and presenting data in a meaningful manner, and inferential statistics, which entail drawing conclusions or making predictions about a population based on sample data. The statistical operation type also reflects the type of research. As stated earlier, the topic of this book is not statistics, but I explain various terms and distinctions here in order to avoid conceptual confusion. Descriptive statistical indicators are found through terms of analysis such as *frequency, ratio, mean, median, standard deviation,* and *variance* to represent the population through the sample. On the other hand, types of analysis such as a time series, meant to make predictions about a population, are inferential statistical indicators.

between creativity and aptitude). Causal-comparative research explores whether a naturally occurring intervention (not controlled by the researcher) affects an outcome (e.g., whether environmental noise affects student performance in the classroom).

3. **Intervention studies** entail a certain method or treatment that is expected to affect one or more results. This type of research is conducted to develop an application. For example, with intervention research, it is possible for researchers to evaluate the effectiveness of numerous teaching methods, curriculum models, classroom arrangements, and other efforts to influence the characteristics of individuals or groups. Intervention studies can also contribute to general knowledge by verifying theoretical predictions. The primary method used in intervention research is experiments.

As mentioned previously, qualitative research studies are mostly descriptive on the other hand, associational research studies are quantitative. Theoretical studies are more innovative in scientific research. However, they require a strong conceptual infrastructure. The assumption is up to the researcher. Identifying deficiencies in game theory and designing a new game theory can be stated as an example of theoretical study.

Analytical research seeks an answer to the question of "**why**." Research can be carried out through reasoning[3] or through experimentation or observation in these types of studies associated with cause-effect relationships (Yang, 2001). The findings from this type of study can be obtained by entering the field to collect data via experimentation or observation or by using computer simulations.

CONCLUDING REMARKS

The methods used in social sciences vary widely and sometimes require the use of certain field-specific methods, single or mixed. In the scientific world, researchers generally use a few methods more widely in their studies. However, most of the time, the key to progress depends on pushing one's limits. When a researcher decides to try a method that she/he has not experienced before (provided that it is suitable for the research topic and question) for the new subject she/he wants to study, she/he usually encounters new difficulties other than the research difficulties she/he is used to and gains new experiences in an effort to solve them. This is a good way for a researcher to improve her/himself.

---

[3] The process of drawing conclusions from one to prove the other of two propositions (see Chapter 2 for logical inference).

There are many routes to reach the city center. Have you ever tried taking a different route? In my opinion, try it, and you probably will see new sights.

SUMMARY OF CHAPTER 6

The scientific methodology includes the stages of determining the purpose of the study, identification of keywords, assumptions, scope, and limitations. A researcher should know different research methods when required to choose the correct method or decide on a mix of methods. Because different scientific subjects, purposes, and resources can require different methods to use, or some research studies can necessitate by their nature to use of mixed methods.

Quantitative methods provide making estimations through data collected from the sample. Quantitative studies can be designed as experimental or non-experimental. Surveys, correlational research, causal-comparative research, experimental research, or observations, are quantitative methods used in scientific inquiries.

Qualitative methods provide making in-depth, holistic descriptions of the unit of analysis and the context, and help build concepts and theories. In most cases, the practice of qualitative methods requires more attention because of their complex nature. Ethnographic research, historical research, action research, phenomenological research, grounded theory research, case studies, narrative studies, and critical research are qualitative methods used in scientific inquiries.

Mixed methods combine quantitative and qualitative methods in the same research study. Some research designs can require using mixed methods, or researchers can desire to benefit advantages of both methods.

Particular classifications can be made according to the outcome features or operational features of research studies. *Basic* and *applied research,* as well as *descriptive, associational,* and *intervention studies* are categorized according to their outcomes.

Chapter 7
UNDERSTANDING THE RESEARCH PROCESS

Before starting to read the chapter

The stages of research processes are not separated by absolute boundaries, that is, they have an intricate structure and are intertwined with each other in some respects. Therefore, the scientific research process can only be shown in outline. For example, during the theoretical reading, one can move away from the question that was considered at the beginning of the study. The context of the question may change or be replaced by a completely new question. Such backtracking and intertwining of stages of research are very common.

This chapter introduces *the scientific research process* step by step according to the order in which the stages are required. There are many scientific resources that present the scientific research process by dividing it into different numbers of stages, but all of them have the same logical basis irrespective of the number of stages. This chapter aims to bring readers a better understanding by presenting a few alternative outlines of the scientific research process to bring readers. Furthermore, readers know the *reliability* and *validity* concepts regarding research findings.

THE SCIENTIFIC PROCESS

Scientific research begins with a question. With a preliminary study, the answer to the question is sought in the literature, and if there is no answer found or if it is seen that the sample on which this question is intended to be studied has not been studied before regarding this question, the question is turned into a research hypothesis. The hypothesis is tested after collecting data from the field. Collected data is processed to make an analysis. Conclusions are reached by making inferences. Finally, the conclusion and evaluations are presented. Scientific research studies are prevalently conducted in this frame in social sciences.

A study is based on two fundamental stages. The first phase is the theoretical framework. At this stage, the variables selected in relation to the research subject are presented with a theoretical basis. All previously published theoretical works (periodically, starting from classical studies to the most recent theoretical studies) and empirical studies should be examined. After this comes the data collection and analysis phase, which requires a strong methodological background. The data collection and analysis phase of a study consists of the following three main steps:

- ⇨ Designing the research
- ⇨ Collecting data and determining the appropriate analysis (or test) method
- ⇨ Analyzing the collected data

In more detail, the scientific research process is as follows (Fraenkel, Wallen, and Hyun, 2011; Root, 2007):

- ⇨ Identifying the subject of the research
- ⇨ Defining the research problem and determining the research purpose
- ⇨ Identifying research questions or hypotheses
- ⇨ Building the research model
- ⇨ Identifying keywords (all terms in the research problem or hypothesis must be clearly defined)
- ⇨ Conducting a literature review
- ⇨ Creating a research design-pattern (sample selection, data collection tools [instruments], measurement procedures, and limitations are determined)
- ⇨ Collecting data from the sample (measurement)
- ⇨ Analyzing the information
- ⇨ Evaluating the findings
- ⇨ Presenting recommendations
- ⇨ Reporting

Figure 7.1 presents the scientific research process, including theoretical study, research design, data collection, analysis, and finally conclusion and evaluation stages, indicating the activities needed at each stage.

Figure 7.1. Scientific research process

Research Process

Phases	Action	Notes
Theoretical Study	1. Step: Identifying general research topic/concept 2. Step: Literature review	1. Step: Professional or area of interest 2. Step: Past empirical and theoretical studies
Research Design	3. Step: Defining the research problem; defining the purpose of research study regarding to the problem 4. Step: Defining the research hypothesis or the research question(s); identifying keywords; identifying research type; building research model 5. Step: Defining the population of research and unit of analysis 6. Step: Defining of the sample, scope and limitations 7. Step: Decision for *measurement method* or *data collection method;* decision for *data collection instrument*	3. Step: Through gap identification. After detected a gap in the written literature the purpose of the study put on forward 4. Step: Starting point at asking a question or asserting an assumption to verify or falsify 5. Step: Individuals, groups, geographical units (towns, cities, states etc.), branch of industry, a company, a business unit/units etc. 6. Step: Constraints of time, fund, and other sources 7. Step: Questionnaires, Observation, Experiment, Document review, Interview etc.
Data Collection	8. Step: Measurement of the sample and determination of appropriate analysis (test) method	8. Step: Conveying the measurement on the sample
Analysis	9. Step: Analysis of collected data (or information)	9. Step: Conveying analysis such as estimation of parameters, comparison of groups, regression, correlation analysis on depending upon type of relationship (correlational or causal), factor analysis, description of qualitative data etc.
Conclusion & Evaluation	10. Step: Interpreting of research findings and making inferences	10. Step: Includes making interpretations and inferences about research findings

Determining the Research Topic or Problem

A distinction must be made between the subject of research and the field of study. The field of study refers to the scientific field in which the researcher/professional is directed in line with his/her profession or curiosity. Economics can be a field of study. Similarly, sociology, psychology, or any of the natural sciences (such as physics) is a field of study. However, academics and scientists commonly focus on more specific areas within these fields. They do this to specialize in a particular subfield or subject because, for example, the field we call economics is very broad and is divided into many sub-branches. Even the accumulated scientific knowledge in the fields we have counted increases so rapidly that it becomes almost impossible to follow the developments in just a single field. For this reason, subfield specialization is common in the world of science and academia, just as in professional life. An academic working in economics chooses to specialize in a particular subject of economics, such as *macroeconomics, microeconomics,* or the *history of economics.* Similarly, someone who decides to study physics in the natural sciences must choose from among many fields, such as *astrophysics, geophysics, quantum physics,* or *theoretical physics.*

The subject of a study refers to the search for a subject within the study areas we have mentioned above. Although the complex nature of social events presents social scientists with an abundance of topics, it is not easy today to find a topic on which to conduct an original study, because the purpose is to provide a concrete contribution to the scientific field in question rather than to determine any subject (i.e., to produce an original study).

After the study area is determined, the original subject is chosen. In other words, the subject of study is narrowed down from general to specific, and the initial question/problem arises in this phase (McMillan, 2015). At this stage, the problem is defined and the purpose is put forward according to the subject. A well-written statement of a problem generally indicates the variables examined in the study (Gay, Mills, and Airasian, 2011).

The research problem often points to an encountered difficulty, inefficient yield, or a situation that needs improvement. In short, there is literally a problem. A research problem includes areas of interest for the researcher, conditions that need improvement, challenges to be eliminated, and questions to be answered (Fraenkel, Wallen, and Hyun, 2011).

Note: Young researchers should not study a subject that does not have a theory, especially in the social sciences! Theories provide a strong background for the study and establish the framework, premises, and standards to be followed. Research studies having a theoretical framework present more accurate findings.

Research topics may include:

→ **General**: E.g. social policies, health policy, general health systems, processes, and programs.

→ **Particular**: E.g. the adequacy of the national elder care policy, the adequacy of disability-friendly infrastructure programs of the local government.

Literature Review

The **literature review** refers to the review of scientific publications on a particular topic. The literature review also gives researchers an idea of subjects that require more research (Fraenkel, Wallen, and Hyun, 2011). The literature review begins after the general research problem is introduced, and summarizes and analyzes previous studies on the same problem/question (McMillan, 2015).

Literature sources may be digitally-based publications (journals, books, encyclopedias, monographs, reports) as well as printed (Fraenkel, Wallen, and

Hyun, 2011). A successful literature review covers respected sources of information in the field and begins with the collection of the top-level (and generally the most cited) articles published in peer-reviewed journals, especially in the field of study (Miller, Strang, and Miller, 2010).

A literature review enables a comprehensive understanding of the conceptual framework. At the same time, it is a content analysis and necessitates a careful investigation to systematically collect and organize the findings of the examined papers. A successful conceptual framework necessitates reviewing the most cited and prestigious publications in the field. Additionally, current publications should be reviewed for the latest findings.

The Research Design

The research design is a general strategy for conducting a scientific study, determined by the nature of the hypothesis, the variables involved, and any environmental constraints (Gay, Mills, and Airasian, 2011). The research design refers to the plan and structure of how the research will be carried out (Rogelberg and Olien, 2014).

> **If we knew what it was we were doing, it would not be called research, would it?**
>
> Anonymous

According to some authors, a research study starts with identifying a research topic/idea. According to other authors, it starts with identifying a research problem. In this book, the research design process is described as following the latter. A research design involves the following steps:

- ✓ Determining the research problem and the study's purpose.
- ✓ Identifying the research question or hypothesis.
- ✓ Determining the population, analysis unit, and analysis level.
- ✓ Identifying the sample and limitations of the research.
- ✓ Determining the measurement method and measurement instrument.

I will explain all steps outlined above in the following subsections.

Determining the Purpose

Scientific research can focus on several purposes, such as understanding a social or natural phenomenon, classifying species, solving a problem,[1] or

---

[1] Problems include social, institutional, or natural problems, and more specifically, violence against women, low motivation among employees, and reduced demand for a firm's products.

evaluating the strengths and weaknesses of a program[2] (Patton, 2014). The purpose of a study should be clearly explained, as it guides the research design (Creswell and Creswell, 2017; Maxwell, 2013). The research question or hypothesis is put forward after the problem and purpose are identified.

Identifying the Research Question or Hypothesis

A hypothesis is a formal statement (and proposition that indicates a prediction about a specific event) that designates the relationship between variables (a dependent and an independent variable). A hypothesis is made to clarify the direction of a research problem (Chigbu, 2019; Creswell and Creswell, 2017). An effective study has specific, meaningful, testable hypotheses based on a compelling model (Balas, 2018).

Hypotheses are based on background assumptions and tested by analyzing the collected data. Incorrect background assumptions or wrong research models can cause incorrect hypotheses to be constructed (Muliak, 1986).

Research questions as hypotheses guide decisions about the research design and methods. Research questions help the researcher link the literature review to the kinds of data to be gathered (Bryman, 2007). Hypotheses and research questions are interchangeable. Some researchers prefer to identify hypotheses. However, identifying a hypothesis is not imperative to carry out a study. Research questions can be used instead.

A good research question should be *feasible* (i.e., it should be suitable for resource use). The research question should also be *clear*, which means that most people should agree on what the keywords in the question mean. Research questions should furthermore be *significant*. Significance refers to the study's potential for contributing to society, institutions, or nature. Finally, a research question must be *ethical*, which means that it will not cause physical or psychological damage to human beings, or to the natural or social environment (Fraenkel, Wallen, and Hyun, 2011).

Different formulations of the research question require different methods to answer them. Also, how a question is asked can impact what needs to be done to answer it (Miller, Strang, and Miller, 2010). Research questions are often based on researching a relationship. The term *relationship*, as used in research, refers to the connection or association between two or more characteristics or qualities. Research questions should be *feasible*, *clear*, *significant* (worth investigating), and *ethical*. Further, a research question or hypothesis should involve keywords (Bryman, 2004; Fraenkel, Wallen, and Hyun, 2011).

---

[2] E.g. social and educational programs; more specifically, online educational programs and learning programs supported by artificial intelligence.

Research questions can be formulated as follows:

Descriptive studies (seek to answer *why* and *how*) (*Qualitative research*)
Example 1: *How* is violence defined in society?
Example 2: *Why* has violence against women increased in middle-class families?
Example 3: *How* can students' focus on lessons be enhanced?
Example 4: *Why* did market volume decrease last year?
Example 5: *How* can employee motivation be increased?
Example 6: *What*[3] factors cause students to be distracted in the classroom?
Relational studies (cause-effect, correlation, or difference-test) (*Quantitative or qualitative research or both*)
Example 1: Is there a relationship between a reward system and employee motivation?
Example 2: *How* does a reward system affect employee motivation?
Example 3: Do teachers treat their students differently according to their gender?[4]
Example 4: Does the fact that employees are not treated fairly affect their motivation?[5]
Example 5: *What* is the relationship between intelligence and academic success?
Intervention studies (cause-effect, experimental) (*Quantitative or qualitative research or both)*
Example 1: *How* does peer pressure affect individuals' behavior?
Example 2: *How* does new medical treatment affect cancer patients?

Building the Research Model

Models are greatly simplified as well as useful representations of actual objects, phenomena, and processes in reality. Modeling is used in many areas, such as business (which usually includes processes of manufacturing or service systems),

---

[3] I added this example to show that the way in which questions are formulated is not so restricted in science, despite the explanations and definitions that imply this is so. In other words, there is no standard for using *how* and *why* tags while describing research questions in qualitative studies; it is just a matter of logic to convert a *what* question into a *how* or *why* question.

[4] Since this question requires making comparisons between groups (female and male students), a causal-comparative method is used; a general research approach can be quantitative or use mixed methods (Fraenkel, Wallen, and Hyun, 2011).

[5] A causal-comparative method is used since this question requires making a comparison between groups (who think they are being treated fairly and are not treated fairly at workplace). The general research approach can be quantitative or mixed methods.

math, or physics. Research models help us to understand essential structures and functions in nature and social life. Research or scientific models, as a particular type of model, offer a simplified representation for a better understanding of complex concepts. In other words, a **research model** is a clear statement of how an important part of nature or society works. A research model can be a brief paragraph, diagram, mathematical formula, or physical object. Hence, research models can ignore many less relevant factors. On the other hand, models are not reality itself; they merely reflect certain aspects of reality (Balas, 2018).

Building a research model offers some advantages for the researcher; for example, the complexity is controlled by the researcher. Models are built on available evidence, so the researcher should have a prior idea of what elements should be included in a realistic model. In other words, research models must be consistent with known facts and knowledge derived from past experiments and observations (Balas, 2018).

Models also have certain principles, such as that a research model is constructed before starting measurements or experiments. This condition necessitates reviewing the literature before building a model. Accurate models can be tested from multiple angles, and the results should be confirmatory. Research models are also expected to meet three main qualifications as being representative of reality. First, the content of the research model should contain all the key aspects of a given construct. Second, the research model should be a reasonable imitation of the real world. Finally, the research model should accurately portray what it is supposed to represent (Balas, 2018).

Let us examine an example of a research model in the marketing field. When we want to understand the influence of social media advertising and product quality on brand awareness under the mediating effect of consumer engagement, the following research question arises:

> *"Does social media advertising (independent variable 1) and product quality (independent variable 2) jointly influence brand awareness (dependent variable) through the mediating effect of consumer engagement (mediator variable)?"*

Let us list our variables included in the research question:

→ Independent Variable 1: Social Media Advertising

→ Independent Variable 2: Product Quality

→ Dependent Variable: Brand Awareness

→ Mediator Variable: Consumer Engagement

→ Control Variables: Demographic Factors (e.g., age, gender, income)

Figure 7.2 shows our research model. Blue arrows show the direct influence of independent variables (social media advertising and product quality) on the dependent variable (brand awareness).

Figure 7.2. A research model example

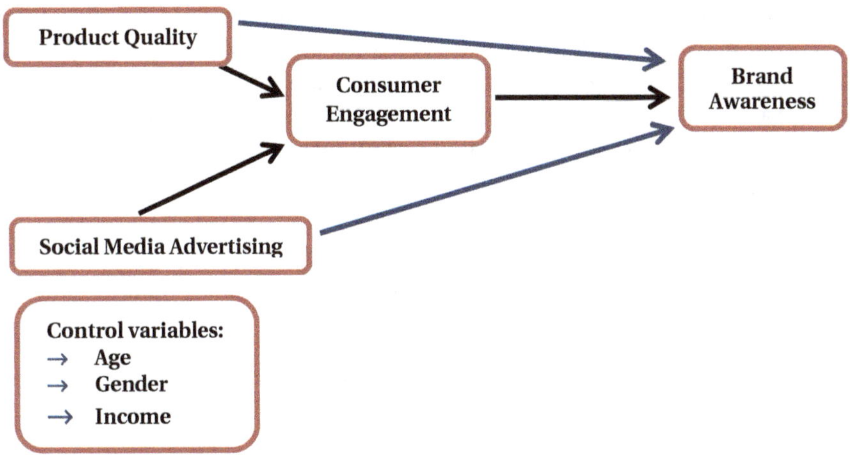

Identifying Keywords

Keywords or phrases are very important components of a study; they primarily help to draw the outline of a study's theoretical framework. Accordingly, it is crucial to conduct a comprehensive literature review in the beginning to identify accurate keywords. At this step, the research problem (or topic) helps the researcher to find related studies; for this reason, it is vital to define the research problem precisely (in accordance with field terminology) (Fraenkel, Walley, and Hyun, 2011).

Keywords also refer to the study's theoretical framework. Thus, researchers may need to re-review the literature to follow the stages of the study. Consequently, the scientific research process, listed stage by stage in this chapter, is not restricted by how the order is arranged.

Measurement of the Sample

Researchers should complete sampling before conducting measurements of the sample. This book does not include sampling methods, which are widely covered in many methodology books. Measurement of the sample is a very technical issue. It may require controlling (or isolating) the research environment to collect data correctly and to obviate factors that can manipulate participants in the environment. In some cases, measurement (e.g., through observation) may need to be conducted in a wholly natural flow of life or in the environment of the research object.

Errors can occur during measurement. One or more variables can be redundant, or sometimes two variables are conceptually distinct but may not be empirically distinct. For example, job satisfaction and organizational commitment are conceptually different but both variables correlate nearly 1.00 after each measure has been appropriately corrected for measurement error (Schmidt, 2010). Moore et al. (2000) show in their investigation that measurement errors are also related to how respondents understand questions and form their answers. Apart from missing data on survey forms, this is another potential measurement error to guard against.

Measurement of the sample is also related to another technical issue, that of reliability and validity, which are the quality indicators of a study. I address these quality indicators in the next subsection.

Reliability and Validity

There are two main quality indicators in research findings: *reliability* and *validity*. Both quality indicators are linked to the research scale and, therefore, to the measurement. **Reliability** denotes the consistency (stability) of the measurement (the obtained scores or responses) of the variable(s) subject to the study, and establishes that the same findings can be obtained when the measurement is repeated. In this regard, it signals that the measurement method is reliable. Reliability means that the measurement instrument has a consistent, stable measurement character. A measurement instrument (scale or test) is reliable if it can gauge a feature with the same outcome every time it is applied. For example, if a newly developed questionnaire produced different results regarding daily alcohol consumption in two separate measurements in the same month (0.25 liters in the first measurement and 0.5 liters in the second measurement), this means that the survey (i.e., scale) is not reliable (Buhringer and Sassen, 2010; Fraenkel, Wallen, and Hyun, 2011; Rogelberg and Olien, 2014).

However, the reliability of a measurement instrument does not lead to validity. For example, let us assume that a new unit of measurement based on 13 mm measurement ranges has been developed. If the intervals are consistent in it, there is reliability, but a 13 mm scale will not be valid because it is not suitable for use. In another example, let us assume that we divide time into slices and produce a clock according to the daily cycle of the planet Venus. This watch will be reliable in determining time, but it is valid only on the planet Venus. In world conditions, our watch would be dysfunctional, and hence will not be valid.

Thus, the reliability of the measurement instrument (scale or test) is not the only criterion. It is also important to consider whether the scale is compatible with the variable (length or time) to be gauged. At this stage, the validity criterion appears. **Validity** means that the findings obtained reflect the variable studied correctly (accurately). In other words, the validity represents what the study was intended to measure, and indicates the structural fit of the variable that is to be measured with the scale. Thus, the data provided by a measurement instrument that passes the validity test are considered appropriate, meaningful, and accurate in the context. In short, reliability is the consistency indicator of the measurement instrument, and validity indicates the quality of compliance of the variable to be gauged with the measurement instrument. Having defects in one or both of these criteria can lead to completely wrong results (Buhringer and Sassen, 2010; Fraenkel, Wallen, and Hyun, 2011; Rogelberg and Olien, 2014). In another example, a reliable unit of measure, such as a foot for length units will give the same outcome for each measurement. However, in a study that examines living things on a microscopic scale, it does not seem appropriate to express the measured values in feet. In this context, microscopic measurements, a foot scale is a reliable—but not valid—scale.

The closer a scale yields measures able to represent the reality of the topic under study, the more valid it is. The more consistent the scale is with previous results (under similar conditions),[6] the more reliable the scale is. These quality criteria are a way of evidencing a study's quality so that the findings can be generalized and accepted as correct. For this reason, measures are aimed to be as *objective, reliable,* and *valid* as possible (Buhringer and Sassen, 2010; Fraenkel, Wallen, and Hyu, 2011).

Reliability concerns the precision of data collection and occurs at the measurement stage. However, validity begins at the phase of sample selection and is a basic quality criterion for all aspects of the study, such as determining the study's questions and research design. In other words, choosing the right sample is the first step of validity. Reliability is indirectly an indicator of *objectivity.* Reliability is essential for a measurement instrument (such as a survey), but is not sufficient in terms of validity. Moreover, reliability is a prerequisite for validity (Buhringer and Sassen, 2010).

The purpose of measurement in research is to measure the real value of a particular variable. The reliability criterion comes from classical test theory, according to which measurements are not perfect and represent a true score as well as a measurement error. The greater the measurement error, the lower the reliability. For this reason, the researcher should try to reduce the measurement

---

[6] A study can be repeated with a new sample and sometimes even under new conditions. This is called **replication**.

error by increasing the reliability of the measurement (Buhringer and Sassen, 2010).

There are two types of validity: *internal* and *external.* A scientific study should meet these two criteria to ensure that it is valid and that its conclusions are trustworthy (Creswell and Creswell, 2017; Mitchell and Jolley, 2009). Internal validity concerns causal factors that exist and can be substantiated and it's related to the *credibility* of the scientific study. External validity is related to its *transferability* (Patton, 2014). In other words, external validity refers to the possibility of making generalizations from the sample to the population. For example, the efficacy of a new drug can be examined for a *specific population of reference* (a subset of a target population, e.g., a participant subset of a certain age group). If the efficacy of the new drug can also be established for *other populations,* it indicates that external validity is provided (Russo, 2017).

External validity problems may cause incorrect inferences from the results (Creswell and Creswell, 2017). *External validity* is important when the main purpose of a scientific study is to provide findings that can be generalized to the real world (generalization to other people, time periods, or situations), since a study providing evidence of external validity aims to show that the participants did not represent a limited sample of people (Mitchell and Jolley, 2009). To ensure *internal validity,* the researcher should adhere to the following steps (Creswell and Creswell, 2017):

- ✓ Collect data from multiple sources.
- ✓ Clarify researcher bias.
- ✓ Make long-term, repeated observations.
- ✓ Promote ongoing dialogue with participants to ensure the true value of the data.
- ✓ Involve participants in all phases of the study, from the research design to the conclusions.
- ✓ Provide peer examination.

Reliability can be statistically estimated. Estimating reliability gives researchers an idea of how much variance they should expect. Such estimates are generally expressed as another application of the correlation coefficient, known as the reliability coefficient. While the validity coefficient indicates the relationship between the scores of the same individuals using two different measurement tools, the reliability coefficient refers to the scores obtained by two people at the same time with the same measurement instrument, or the relationship between the scores obtained from two parts of the same measurement instrument as applied to the same people (Fraenkel, Wallen, and Hyun, 2011).

Reliability is calculated by correlating different measurements. The reliability coefficient takes a value between 0 and 1. In other words, unlike other correlation coefficient tests, it has no negative value. When the measured variance and the actual variance are the same, the reliability is scored as 1 (i.e., the variance of the measurement error is 0). Reliability is expected to be at least 0.8 to be considered acceptable (Buhringer and Sassen, 2010; Fraenkel, Wallen, and Hyun, 2011).

Systematic action is required to increase the reliability and validity of the measurement. Some applications that can be implemented in this direction include the following (Buhringer and Sassen, 2010):

- ✓ Conducting the measurement process by adhering to a set of procedures and performing measurements according to a set of written instructions; the environment, instrument, and subject will accordingly contribute to the uniform execution of the research at each stage.
- ✓ A measurement's reliability can be enhanced by training the researcher (or observer) to apply the measurement in the same manner (i.e., in a uniform way).
- ✓ The measurement instrument needs to be refined and written to prevent misunderstanding. In addition, the questionnaire or interview questions should be clear.
- ✓ Automating the measurement instrument can reduce variability in the researcher's way of taking measurements.
- ✓ By repeating the measurement, the effect of random error is reduced and the reliability of the results obtained is increased.
- ✓ If it is possible to make the act of measurement not apparent, the measurement results cannot be deliberately directed by the subject.
- ✓ Repeat adjustment of the measurement instrument increases the validity of the measurement so that the intended aspect can be gauged.

Analyzing the Data

The data gathered before the analysis should be scored accurately and consistently. After scoring, the data should be tabulated and coded. This makes it possible to group the different types of data collected, in addition to demographic data, to compare and analyze different datasets in various ways

(Fraenkel, Wallen, and Hyun, 2011). Table 7.1 shows various packages and their functions.

Table 7.1. Various analysis package programs

<table>
<tr><th>Program</th><th>Functions</th></tr>
<tr><td>Spss
Excel
Minitab
Sas</td><td>✓ T tests (Used to compare the means of two groups)
✓ Z test (A statistical test used to determine whether two population means are different when the variances are known and the sample size is large)
✓ Chi-square tests (Used to determine if there is a significant association between two categorical variables)
✓ Normality tests (Used to check if a dataset follows a normal distribution)
✓ ANOVA (Analysis of Variance: Used to compare means across three or more groups)
✓ Correlation analysis (Used to measure the strength and direction of the linear relationship between two variables)
✓ Regression analysis (Used to analyze the relationship between one dependent variable and one or more independent variables)
✓ Non-parametric tests (Statistical tests that do not require the assumption of a specific distribution for the data)</td></tr>
<tr><td>Lisrel/Amos</td><td>✓ Structural Equation Modeling (SEM)</td></tr>
<tr><td>Stata</td><td>✓ Preferred for panel and time series regression analysis and facilitates working with large data sets</td></tr>
<tr><td>Fsqca</td><td>✓ Fuzzy-set Qualitative Comparative Analysis</td></tr>
<tr><td>Ucinet
Vosviewer
Gephi
Cytoscape</td><td>✓ Network analysis</td></tr>
</table>

Drawing Conclusions and Evaluating the Findings

The conclusion is an overall summary of a study regarding whether the hypothesis is supported (McMillan, 2015). In other words, all findings are discussed in the conclusion in terms of the research problem or hypothesis. The conclusion also presents generalizations consistent with the results. Hence, all evidence can be reviewed briefly in the conclusion, which may also contain suggestions for future studies (Gay, Mills, and Airasian, 2011).

Researchers may discover contradictions between research findings and the theory (and its general hypothesis) that the study is based on. According to Popper (1972), contradictions can be fruitful for scientific advancement in each field (Ultee, 2015). For this reason, it might be helpful to state contradictions in the findings (placed in the conclusion) to better comprehend the subject or problem.

Another important point is that while researchers refer to other studies related to the subject for the findings of the research they carry out, they should not ignore studies that contradict their findings. It is an ethical problem (also see Scientific Ethics) when the researcher knowingly does not include other findings that contradict his/her own. If possible, contradictions should be provided with a logical explanation in light of the findings of scientific studies. If a logical and concrete explanation cannot be reached, a contradiction between the current study and other studies may be left open as a question for future studies.

Reporting

A research report includes a theoretical framework and research components. The theoretical framework must include everything directly related to the research subject and purpose, and exclude anything irrelevant. On the other hand, researchers sometimes experience difficulties on whether to include certain details or explanations that are indirectly relevant to the subject. There can be blurry areas that require explanations or definitions. In such cases, researchers may place these definitions (which are not necessary to explain in the text directly) into footnotes.

The final report should be proofread at least twice. In this phase, researchers may ask a friend to review and critique the manuscript for them (Gay, Mills, and Airasian, 2011; Katz, 2009). Finally, all reports need to be edited (Gay, Mills, and Airasian, 2011).

CONCLUDING REMARKS

All activities, from determining the research problem to reporting the findings are part of the research process. Each stage requires studying carefully and elaborately. For this reason, it would be beneficial, especially for young researchers, to read more about the subject or consult other researchers at any stage when they hesitate (and the feeling of hesitation is not a bad thing). There is no problem in moving slowly in research, the main thing is to avoid mistakes.

SUMMARY OF CHAPTER 7

Scientists start their study by determining the subject they want to research. After determining the general subject, they start a literature review. Sometimes literature reviewing can even be the first step when researchers have difficulties finding a general subject. After the determination of the general subject, it has to be narrowed down to decide the particular subject. A literature review is an important step because the gap(s) in the literature can only be seen by comprehensive browsing.

When a researcher has decided on a particular subject the next phase is the definition of the research problem and the determination of the purpose. Always keep in mind before the definition of the research problem and determination of the research purpose, a researcher first needs to do an intensive literature review! In other words, research design can be started only after a comprehensive theoretical study.

Research design starts with the *identification of the research problem.* Identifying keywords is related to the specific subject, problem, and variables of the research study. Keywords should be derived in accordance with field terminology. Then, the purpose of the research is determined. Research questions or hypotheses arise related research problem and purpose of the study.

A *research model* is an illustration representing the interaction/relations between elements/variables. Models help us to understand a structure, a system, or functions.

After deciding the research approach (type), researchers define the *population* and *sample.* In this phase, a researcher should also determine the limitations of the study. Measurement of the sample starts after the determination of the *data collection method* and *data collection instrument.* Measurement of the sample is the data collection phase of the research process.

The latest phases consist of the *analysis of collected data,* conclusion and evaluation of findings. The conclusion is an overall summary of a study. A researcher should state contradictions in the findings as well and strive to explain the possible reasons via the findings of other studies in the literature.

Chapter 8

SPECIAL TOPICS

Before starting to read the chapter

In scientific studies, sometimes we may not realize where the problem originates or what the deficiency is. For example, a good research study and rich findings can sometimes have less impact than they deserve if not presented well. The problem I point out here is related to the presentation of the findings. Findings are now presented visually, with an increasing trend. There are many programs that help with this, and it is even possible to get artificial intelligence support. A successful visual presentation can make even a mediocre work appreciated (do not let your work be mediocre, however).

The war between science and ignorance (if there is such a war) seems like an -infinity war.- The reason why we will include this subject in this book points to the fact that scientists cannot be isolated from the attribute of ignorance. Because, contrary to what is thought, ignorance is not a general feature, but the fact is that there are many subjects on which every person is ignorant. I also want to point out that the opposite of ignorance here is the virtue of self-knowledge. This is my personal opinion and I know it looks so philosophical. But my point is that a person can be free from this negative adjective (despite all his ignorance) in the case of self-knowledge.

This chapter aims readers to bring an understanding of the relation between *science and ignorance.* They can also recognize and distinguish *types of deception.* Furthermore, after this chapter, readers can detect *pseudoscience,* and its types and activities. Readers are also anticipated to understand *the importance of visualization* in the presentation of data and information.

This chapter initially starts with special topics regarding scientific thought. The first one of those special topics is *science* and *ignorance.* This sub-section is intended for providing readers with an awareness of the word ignorance and its relation to science. Another sub-section in this chapter is the *art of deception.* It is a special topic, and I believe that a scientist or researcher should recognize the types of deception to avoid her/himself both manipulating people/ data and being manipulated by others through tools of deception. The third subheading in Chapter 8 is *pseudoscience,* which discusses the science-looking activities. *Methodological errors,* which can occur in several stages of research studies, are given wide coverage in this chapter. The last sub-section is *visualizing information,* which is a very significant quality indicator of a well-structured presentation of a research study

SCIENCE and IGNORANCE

We humans do not have detailed knowledge about the extent of our ignorance, given the nature of things. Moreover, according to many philosophers, our conceptions of things are always *corrigible, incomplete,* and *changeable* (Rescher, 2003). How can we know whether our knowledge of things about the world is correct? Does our knowledge about things reflect reality? Would we need verification of our knowledge about the world in some way? Or do we need some falsification of our knowledge? Science can function both ways. In other words, science justifies or falsifies a hypothesis. According to many scientists and philosophers, falsification is a natural part of scientific endeavor in an epistemological sense (Reed, 2012; Rescher, 2003).

Data, evidence, and rational considerations shape knowledge (Creswell and Creswell, 2017). Science is about knowledge and the search for it. Undoubtedly, an individual, including those in the scientific community, cannot possess comprehensive knowledge encompassing every aspect of our world.

Ignorance has many definitions. In an academic sense, ignorance refers to a lack of knowledge or awareness about a particular subject or topic, often accompanied by a disregard for seeking or accepting relevant information. Ignorance can manifest in various forms, including factual inaccuracies, misunderstandings, misconceptions, and biases (Proctor and Schiebinger, 2008; Smithson, 1985). Moreover, there are many notions that can be associated with ignorance, such as false belief, error, absence of knowledge, and lack of true beliefs, doubt, and misinformation (Arfini and Magnani, 2020). Everybody is ignorant of something. The word "*something*" is intentionally used. Each person is ignorant of many things. If someone specializes in finance, she/he can be consulted for her/his expertise and knowledge related to finance. However, the vast majority of finance experts have no idea how to cultivate the land. Similarly, scientists specialize in a particular field of science. Thus, they are trustworthy in their specialty yet, they are open to making mistakes.

People's ignorance is often invisible to them. They remain ignorant of the many ways in which they demonstrate gaps in knowledge. People are often successful in determining a few areas where their knowledge and expertise are lacking. Notwithstanding, an individual's mental catalogue of her/his areas of ignorance can be considered much wider. Ignorance is strengthened by the condition of not knowing what people do not know. In other words, ignorance lies in the realm of *unknown unknowns.* Ignorance is also related to a lack of self-insight into ignorance (Dunning, 2011).

Unfortunately, ignorance can be fatal for both individuals and societies. Fake science narratives (like fake news) undermine the credibility of science and the capacity of individuals and society to make evidence-based choices (whether democratic or individual). For example, the *vaccine hesitancy* movement caused record levels of measles infections in Europe, despite the proven record of the efficacy of vaccines in preventing infections from deadly diseases. In another instance, the denial of anthropogenic climate change has resulted in international agreements on climate change losing universal acceptance. We will probably experience the bitter consequences of global warming on the worldwide level in the 21st century (Hopf et al., 2019).

> **"The saddest aspect of life right now is that science gathers knowledge faster than society gathers wisdom."**
> (Isaac Asimov 1988)

One of the most famous and frequently cited theories about ignorance is the Dunning-Kruger effect, which asserts that people with substantial deficits in their knowledge or expertise are not able to recognize those deficits. They are overconfident, despite making error after error. They tend to think they are doing just fine. They are incompetent and have little insight into their incompetence. The ignorance problem imposes a double burden. First, people's deficits in their expertise/knowledge lead them to make many mistakes. Second, those same deficits render them unable to recognize their own mistakes, or to understand why other people's responses (or choices) are wiser (Dunning, 2011; Kruger and Dunning, 1999).

Science consists of activities that seek answers to unknowns. The problem starts right here; what are the unknowns? How do we know what we do not know? Unknown unknowns are problematic for science. For this reason, researchers conduct their investigations based on known unknowns. Researchers develop hypotheses to test the null hypothesis and have an idea about the results that can reach within a range of known possibilities. While the discovery of one of the previously unknowns shows us how little we know and also presents a fruitful area for new studies. Researchers can examine this newly known under traditional hypothesis forming and testing (Logan, 2009). Thus, it is very important for researchers to consider the possibility of unknowns and uncertainties in their scientific inquiry.

THE ART OF DECEPTION

Under this title, I address the art of deception from a broader perspective in a philosophical and controversial context beyond the methodology context. For a researcher or scientist, it is necessary to be aware of all kinds of deception. A researcher or scientist should not only be aware of scientific deception, but should also generally be aware of how we encounter the art of deception in everyday life, as well as able to draw boundaries to protect the logic of science, thereby protecting their fields of study against deception. For this reason, I will explain all types of deception with examples to raise awareness under this heading.

The art of deception involves convincing people of a reality that does not exist; it is based on the production and sharing of false information. The art of deception can occur through methods such as *lies*, *manipulation* (misleading someone, perception management, or perception operation, which politicians use), *incomplete (or missing) information transfer*, *exaggeration*, *demagogy*, *propaganda* (and *asymmetric propaganda*), and *ad-hominem* statements (Buller and Burgoon, 1996; Ekman, 2009; Vrij, 2008).

The simplest form of deception is a lie. There is an anonymous well-known quote: "*There are three kinds of lies: lies, damned lies, and statistics.*"[1] Of course, do not use statistics to produce a lie! This quote is just information to warn you!

Let us look at another quote from a mini TV-series by HBO titled *Chernobyl* (2019): "*Every lie we tell incurs a debt to the truth. Sooner or later, that debt is paid.*"

In the same series, the following quote is mentioned: "*The truth does not care about our needs or wants—it does not care about our governments, our ideologies, our religions.*"

The art of deception can sometimes be used in complex ways, such as *partially conveying a truth* (omission-excluding part of the information that changes the interpretation of the whole) or *diluting the subject* with false and irrelevant information and arguments.

Visuality is another form of deception. In ancient times, charlatans or demonstrators were able to make people believe in miracles through their abilities at *illusion*. For *optical illusions*, let us explore the example of the Ames room (Ames demonstrations). It is a distorted room that creates an optical illusion (Day, 1993). Today, false information can be produced visually through *Photoshop*, *visual assembly*, or any kind of *visual computer program* or with the help of *artificial intelligence*.

Sometimes, deception is possible with the help of professionally prepared visual tools that give a scientific impression. For example, bar charts should be exhibited at the right height in accordance with data values. But in some cases (e.g. sometimes can be seen in mass media or professional reports), visualization of data does not correspond to the facts. It means that the messages conveyed in visualizations can be completely altered through preferences in how data are

---

[1] The quote has been mistakenly attributed to Benjamin Disraeli (1804–1881), the British Prime Minister, or the famous American author Mark Twain (1835-1910). The origin of the quote is controversial.

aggregated. These visual deceptions are scarce but can be made mistakenly or intentionally (Correll and Heer, 2017). Those explanations are a warning to young researchers!

The deliberate *misrepresentation* and *misinterpretation* of the findings of scientific research is another form of deception. There are studies in the history of science that distort research findings with career or ideological concerns. These are major ethical problems.

There are adherents of the non-scientific side of particular subjects that are publicly well-known, such as deniers of climate-change, the anti-vaccination movement, flat earthers, and deniers of evolution. Unfortunately, these movements have always found allies, albeit in small numbers, in the scientific world.

Another concept related to deception is *bias*. It is easy to sneak conclusions (politically or scientifically) into a mind full of bias. Ethnic codifying (or blacklisting) is still prevalent in social media and even mass media. Most labeling is rooted in bias about Arabs, Jews, Blacks, Gypsies, Indians, and homosexuals. The list can be extended. Some of them are ethnic labeling; some are cultural or religious labeling. I would like to provide a particular warning to readers regarding this problem, since it is vital for a researcher (especially when studying ethnographic topics, which, of course, cannot be restricted to this field of science) to be aware of this kind of bias, which relies on easily generalizing a negative feature of an ethnic or cultural group.

Sexist bias is another problem not only targeting women but also homosexuals. It is possible to find traces of sexist thought codes in almost every society in the world. While a scientist has the duty to expose these biases (via deterministic, descriptive studies), she/he can also be expected to offer suggestions on how to overcome them, because, let us remember that the function of science is not just to identify problems but also to help us solve our problems. If sexism and all other types of bias are a threat to social peace and justice, which they are, then there is a problem that needs to be solved.

The technical aspects of deception, related to research methods, are also important. As will be remembered, deception occurs by hiding facts about research from participants. In Chapter 2, I stated that the problem is found in most research studies, as it is not always possible to share the entire content and purpose of a study with the participants in detail. In this context, the contribution that the participant can provide may be damaged, or it may arouse mental ambiguity in the participant regarding the content or purpose of the study. On the other hand, intentionally lying or misleading the participant about an important component of a study is another form of deception that can invalidate the results (Cunningham, Weathington, and Pittenger, 2013).

In scientific studies, deception sometimes occurs for different purposes. May (2020) outlined four main goals of scientific research, as follows: (1) knowledge production, (2) promoting the acceptance of an ideology, (3) gaining credit, and (4) profit. Except for the first, these motivations of engaging in science may display potential problems in terms of scientific ethics, since they have the potential to misrepresent data for private benefit.

PSEUDOSCIENCE or FAKE SCIENCE

Science always seeks evidence to verify a proposition. A proposition cannot be affirmed scientifically without evidence to confirm it. In most cases, one or a few pieces of evidence may not be enough to verify a proposition. Usually, the limited number of studies and the evidence they present show only the possibility of the correctness of a proposition. In some cases, affirming a proposition may be required, so that many studies can be conducted under the same conditions with previous studies or under different conditions. Additionally, as for extraordinary propositions, as Carl Sagan once stated, "*Extraordinary claims require extraordinary evidence.*" This is a very useful motto to keep in mind. Now, let us take a quick glance at fake sciences.

Astrology is a pseudoscience (Allum, 2011); its claims are unfounded. Astrology uses resemblance thinking (e.g., Mars is red; red is the color of blood, therefore, Mars rules war and anger; Mousseau, 2003). Horoscope analysis is a practice carried out within astrology; it is related to a person's sky map at the time of his/her birth (Tseng et al., 2014).

I do not believe in astrology; I'm a Sagittarian and we're skeptical.
Anonymous

Numerology is another example of a pseudoscience that mostly follows absurd ideas and practices (Wallis, 1985). **Parapsychology** (Marshall, 2012; Thagard, 1978), **magic** (Settle, 1971), **fortune-telling, magnetic therapy, feng shui,** and **lunar effects** (Tseng et al., 2014) are also examples of pseudoscience.

Marshall (2012) suggested four indicators to distinguish whether something is pseudoscience. First, it may be fraudulent or deliberately manipulative, trying to pass itself off as science. Second, the underlying theories may be far-fetched or fundamentally implausible. Sometimes it may involve a method, approach, or general mindset that "appeals to observation and experiment," but which "nevertheless does not come up to scientific standards." Even the treatment appears to be pseudoscientific. The fourth indicator to detect pseudoscience is the usage of *post hoc* analysis.[2]

Pseudoscience may benefit from scientific methods, as mentioned above. However, efforts to use pseudoscience for falsification may be based on ideological concerns (e.g., anti-evolution movements). Pseudoscience does not work while applying the standards of *predictability, falsifiability,* or *reproducibility.* These criteria are essential for science. Any field of science has evidence-based theories, as well as scientific laws proven by experiments, observations, and surveys. Scientific theories display *internal consistency* and *epistemic norms.* Further, science always requires *scientism.*

METHODOLOGICAL ERRORS

We should remind that again researchers should identify validity threats, such as their own biases and assumptions, as well as flaws in logic or methods in every stage of scientific inquiry (Maxwell, 2013).

Measurement error expresses the differences in scores obtained from the same individuals with the same measurement tool. There are two types of measurement errors that can be categorized by their source as *bias* (systematic) and *random* errors. First, **bias errors** occur due to a feature of the survey design, such as *unclear definitions, poorly worded questions, requesting the data from the wrong person* in a business, and/or *data capture problems.* Bias errors produce results that favor one direction. In other words, bias errors are a consistent measurement error that reduces the validity of a statistical test due to pushing scores in a certain direction. In contrast, **random errors** (deterministic errors) do not favor any direction of results. The findings stem from the unsystematic random error of measurement. Random errors display probability distributions with their own variance (Fraenkel, Wallen, and Hyun, 2011; Mitchell and Jolley, 2009; Snijkers et al., 2013).

In some cases, **theory-related bias errors** occur. Believing that the asserted hypothesis must be correct before all evidence is gathered may cause the researcher to interpret the evidence selectively. Similarly, paradigms can become biased and paralyze researchers (McKnight, 2011). Researchers may tend to keep using one method when they suppose they are successful at using it. This methodological choice can prevent them from using other techniques (McKnight, 2011).

---

[2] **Post hoc** analysis is a wrong approach to logical inferencing. It simply means that one event comes after another event, so the first event is the cause. The sequence of events is not the only standard for cause and effect.

Another bias error is common method bias, which indicates a measurement error. **Common method bias** represents a form of measurement error arising when respondents answer questions about multiple variables (likely to correlate) within the same survey. Respondents may tend to provide responses that correlate with each other to avoid cognitive dissonance. Thus, a shared common method may produce a systematic effect (bias effect) on the observed correlation between the variables. For instance, if a survey measures both job satisfaction and job commitment, respondents may answer in a manner that aligns their job commitment answers in coherence with answers to job satisfaction rather than providing independent assessments of each construct. Therefore, the survey wouldn't reflect true scores about job satisfaction and job commitment (Kamakura, 2010: 23). But it's important to clarify that this error arises from the method of measurement rather than the instrument itself being faulty. It should also be noted that common method bias does not always stem from avoiding cognitive dissonance. It may sometimes arise in the context of the participant's particular situation, for example, it may arise from the participant's anxiety about social acceptance.

The scientific world witnessed many errors in the history of science. Some researchers, such as Giora Hon (1989); Onwuegbuzie and Daniel (2003); and Allchin (2012), categorized these errors. In Table 8.1 one such categorization of errors in science done by Allchin (2012) is shown. He also proposes methods against the various errors.

Table 8.1. Errors in Science and developed methods against those errors[3]

Error	Method
Placebo effect	Blind clinical trial
Observer effect	Double-blind
Coincident variables	Controlled experiment
Sampling error	Statistical analysis
Biased sampling	Randomization
Instrument malfunction	Calibration
Inappropriate inferences from results and use of methods	Peer review
Gender or cultural bias	Communal checks and balances

[3] Source: Adapted from the table created by Allchin (2012)

Allchin (2012) also indicate taxonomy of error types as follows;

- Incorrect or impure sample
- Improper procedures such as violation of experimental protocol, poor technical skill
- Insufficient controls to establish domain of data or observations
- Observer perceptual bias
- Sampling errors such as statistical rarity, weak significance level cutoff, or other probabilistic factors
- Flaw in reasoning
- Cognitive biases such as normalizing
- Theory-based cognitive bias,
- Sociocultural biases
- Improper statistical model
- Improper modeling
- Miss-specified assumptions or boundary conditions
- Communication failures such as incomplete reporting, obscure publication, translation hurdles, patchy citation, and patchy search system
- Public misconception of scientific results because of poor science education and poor science journalism

Onwuegbuzie and Daniel (2003) introduce analytic and interpretational errors. The authors indicate qualitative research has the potential for the following general analytical and interpretational errors:

- Failure to legitimize research findings
- Generalizing findings beyond sample
- Failure to estimate and interpret the effect sizes

The authors also indicate general errors in quantitative research as follows:

- Mortality: Researchers delete some data sets for several reasons (e.g., to make equal sample sizes of groups). Such removal can lead to analytical errors if the removed data represents valid cases or different cases.
- Non-interaction-seeking bias: This type of error occurs when researchers do not examine the presence of interactions in their research model. This error mostly occurs in correlational-based analyses such as

correlations, regression, canonical correlation, path analysis, and structural equation modeling. Non-interaction-seeking bias leads both to errors at the data analysis stage and interpretational errors.

- Researcher bias: Occurring at the data analysis stage, it is also known as the *halo effect*. When researchers allow their prior knowledge of the participants to influence the scores given and data then becomes biased. This error also leads to analytical errors.
- Treatment replication error (McMillan, 1999; Onwuegbuzie and Daniel, 2003): Correctly identified units of analysis are essential for research studies. For example, to appoint an individual level of analysis is an analytical error when treatment is received by a group and vice versa. In other words, this error occurs when researchers use individuals as the unit of analysis to compare groups or when researchers use groups as the unit of analysis to compare individuals.
- Violated assumptions: This error occurs when researchers do not consider the underlying assumptions of a particular type of statistical test.
- Multi-collinearity: This error occurs when two or more independent variables are highly related. Multicollinearity is often associated with linear analysis.
- Miss-specification error: It occurs when researchers omit one or more important variables from a research model. This error derives from a weak or non-existent theoretical framework.
- Effect size: It is a common interpretational error, and non-reporting of effect sizes often culminate in the misinterpretation of p-values.
- Confirmation bias: This error indicates a tendency for interpretations and conclusions based on current data to be overly consistent with a priori hypotheses. In other words, available theories might dominate research in a way that blinds the researcher to potentially informative observations.
- Distorted graphics
- Illusory correlation (Kundu, 2015; Onwuegbuzie and Daniel, 2003): It indicates a tendency to overestimate relationships among variables that are either not related or only slightly related. A relationship or association between two or more variables does not necessarily indicate a correlational relationship. Sometimes it can only show the similarity or difference in the movements of two variables. In such cases, it is necessary to interpret the results carefully, specifying the type of relationship obtained. For example, there is an increase in the consumption of ice cream, with cases

of drowning in the summer. Both obviously increase under seasonal influence. However, the high correlation observed in this situation does not imply that there is a relationship between drowning events and ice cream consumption.

- Crud factor: When researchers use large sample size, many trivial relationships can emerge as being statistically significant because, to some degree, "everything correlates to some extent with everything else". The tendency to reject null hypotheses when the true relationships are trivial is called the crud factor.

Using human standards when examining animals in ethological studies can lead to analytical and interpretative errors. For example, it would be biased for a research study to use human intelligence or behaviors as a criterion when studying animal intelligence or behaviors. On the other hand, standardization is important for providing objectivity in scientific studies in some cases. In other words, scientific fields can particularly require standardization. For example, the categorization[4, 5] of diseases is used in medicine to prevent subjective judgments about the states of potential patients. Briefly, researchers should know field-specific methodological errors (as they should follow field-specific ethical standards).

VISUALIZING INFORMATION

The **visualization of information** can be defined as the visual presentation of information that has been systematically collected. Data visualization is not used only by researchers and statisticians, but also by professionals in different fields (D'ignazio, 2019).

The visualization of information is often used in two main ways. First, it is employed to aid *analysis*, and second, as a tool for *communication* (Evergreen and Metzner, 2013; D'ignazio, 2019). The visual organization of data facilitates analysis and evaluation. Scientifically, when a dataset (information) is organized visually, it becomes easier to present the information or to draw inferences from it. Tables, pie or bar charts, line graphs, stream graphs, networks, distribution maps, tree maps, chord diagrams, word clouds, infographics, illustrations, and running animation charts can be used to visualize information (Azzam et al., 2013; D'ignazio, 2019; Rivera, 2020). Various types of charts can be used in editing data. Figure 8.1 shows types of charts.

[4] E.g. The International Classification of Diseases (ICD) is developed by the World Health Organization (WHO).

[5] E.g. The Diagnostic and Statistical Manual of Mental Disorders (DSM) is developed by the American Psychiatric Association (APA).

Figure 8.1. Types of charts

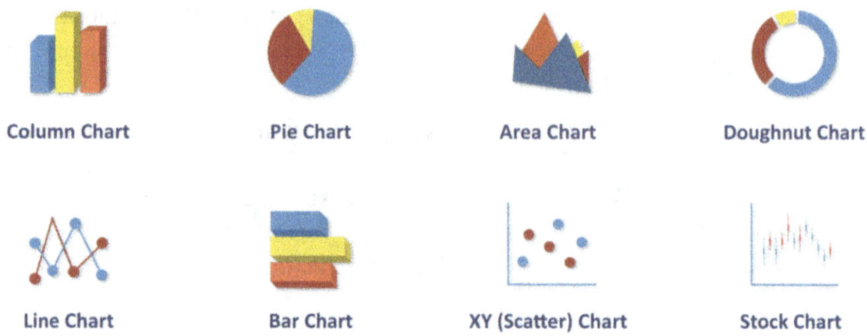

Each chart suits a particular purpose. For example, pie charts and doughnut charts can be used to represent a whole made up of parts. A line chart is mostly used to exhibit data in a timeline.

The visualization of information helps us gain insight from large, complicated datasets that would be impossible to understand in plain text or large tabular form. Data visualization necessitates a sense of color, contrast, balance, and alignment (D'ignazio, 2019). It is also very important to visualize data objectively and within clean layouts (Kennedy et al., 2016).

A use of simpler visual elements also lays the groundwork for emphasizing selected elements to draw the viewer's attention to areas of greater importance. Stripping away the unnecessary and deemphasizing supporting information simplifies the data display, so that an abundance of visual elements does not distract viewers. Secondary data should be simplified to a shade of gray so that chosen elements can appropriately stand out when selected emphatic colors are applied (Evergreen and Metzner, 2013). When deciding on the visualization of data, researchers should pay attention to the following (Franconeri, Padilla, Shah, Zacks, and Hullman, 2021):

- ✓ Visualization design should be coherent with the data type
- ✓ Visual grouping is useful to control a set of comparisons
- ✓ Highlighting is a useful tool to emphasize the single most important comparison
- ✓ A researcher intending to visualize data should know common visual illusions and confusions
- ✓ Supporting understanding is especially important when the intended audience may have low knowledge about the field
- ✓ Visualization should be made in accordance with the audience if the target audience has limited knowledge about graphing and numerical conventions Researchers should learn new visualization formats

Figure 8.2 shows two alternative graph representations for the same data.

Figure 8.2. Percentage of age 65 or older people receiving income by source[6]

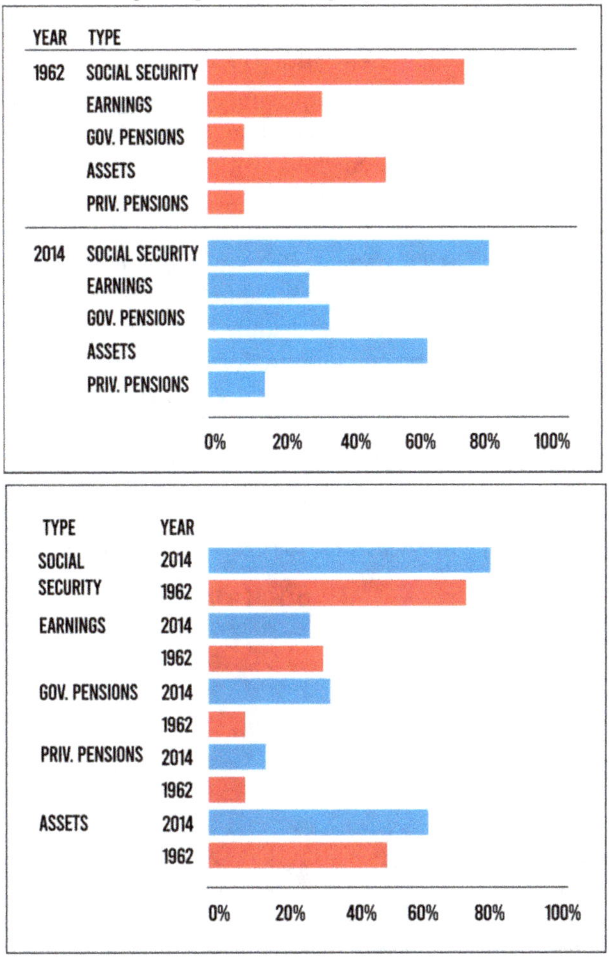

Both graphic representations are correct. While in the chart on the left, the data is grouped by year, in the chart on the right the data is grouped by source of income. In the chart on the right, it seems easier to compare years by source of income. In the chart on the left, it is much harder to accomplish such a comparison since the pillars representing the same source of income are drawn apart from each other. Thus it requires much effort to accomplish a comparison for each source of income. From this point of view, the chart on the right can be assumed to be more accurate for a comparative presentation of data. Briefly,

[6] Source: Recreated from original source created by Franconeri et al., (2021). Data source: https://www.ssa.gov/policy/docs/chartbooks/fast_facts/2016/fast_facts16.html#page5

sometimes data presentation can be a hard task and requires recognizing details while using charts.

Infographics are another way of visualizing information. Infographics are simple communication tools comprising several graphs, diagrams, or illustrations related to a single topic, and are mostly presented on one page. Infographics are used for communication with external stakeholders, and also in education. Infographics are easy to read and promote at-a-glance comprehension (Azzam et al., 2013; Davidson, 2014). They integrate various graphical elements such as illustrations, charts, and graphs to convey information in a concise and accessible manner. By combining visual and textual elements, infographics aim to enhance understanding, facilitate comprehension, and engage audiences in exploring and interpreting the presented content. Effective infographics leverage design principles and visual storytelling techniques to communicate messages effectively and efficiently across different contexts and audiences (Cairo, 2016; Segel and Heer, 2010; Tufte, 2001). Figures 8.3, 8.4, and 8.5 exhibit several examples of infographics.

Figure 8.3. Infographic presentation: 10 largest populated cities in the world[7]

A data set can be presented differently as infographics as presented in Figure 8.3, which actually presents a column chart. The difference of the figure is that the columns have been replaced by structures that are the symbols of the cities.

[7] Data source: https://worldpopulationreview.com/world-cities

Figure 8.4. Infographic presentation: 10 largest countries in the world, 2020[8]

Figure 8.4 presents a data set on the world map. This kind of data set is also proper for being exhibited in the column chart. However, contingent upon its intended usage, displaying the information on a map might be deemed more suitable. What necessitates consideration here is the specific purpose for which the visual aid was created and determining the most fitting mode of presentation accordingly.

Figure 8.5. Infographic presentation: what do 7 billion people do in the world?[9]

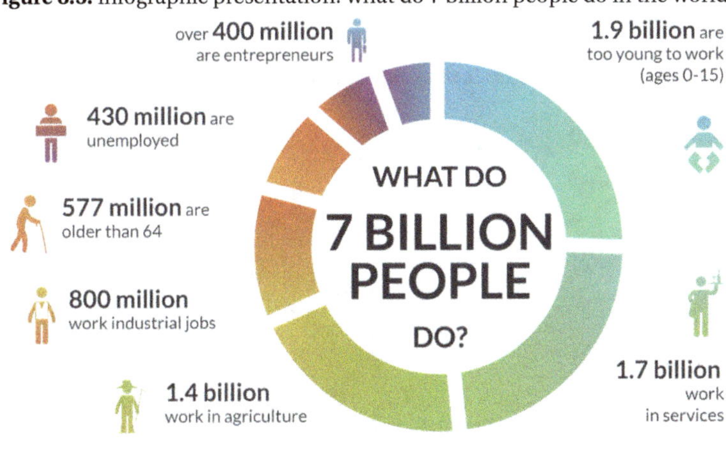

[8] Data source: United Nations World Population Prospects, 2019

[9] Image source: Recreated from original source created by Anna Vital

Figure 8.5 presents a data set on the world population. This kind of data set is also proper for being exhibited in the pie chart. Pie charts are universally regarded as a suitable means of visually representing the constituent components comprising an entirety, signifying either 100% or a statistical value of 1.00.

Figure 8.6. An example of a social network graph

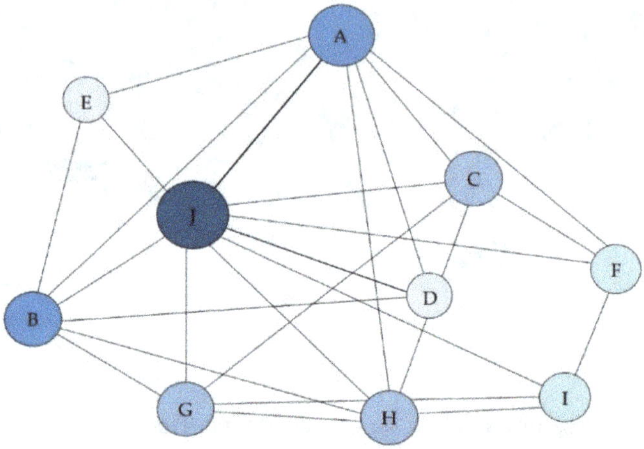

A **network** is a linked set of units such as people, institutions, computers, objects, etc. Networks are systems consisting of units and relationships/ connections between those units. Networks have two fundamental components; *nodes* and *edges*. Node is an ensemble of connected actors. Edge is the relationship/connection between nodes (Golbeck, 2013). These connections can stem from any reason, such as sharing books between students, collaboration among institutions, etc.

Figure 8.7. An example of the betweenness centrality of social network

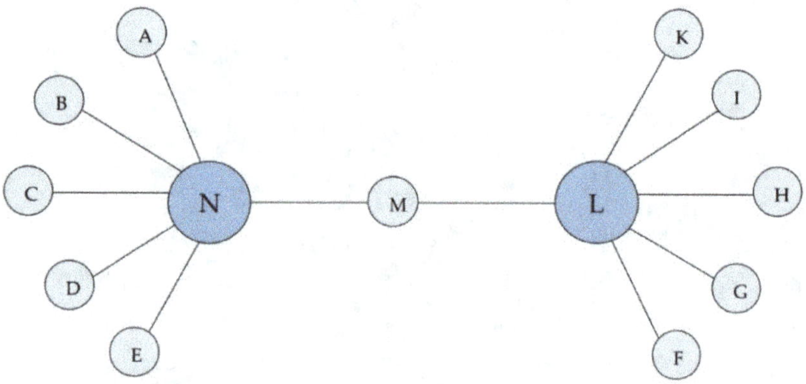

There are some other concepts associated with social network analysis, such as *degree* (centrality), *closeness*, and *betweenness*, the *strength of weak ties*, *structural holes*, and *network density*. Degree refers to the number of connections a node has. "The higher the degree, the more central the node is" (Golbeck, 2013: 27). Figure 8.6 exhibits a visualization of the friendship graph based on degree distribution. In the figure, the actor size and color represent a particular node's number of connections; a larger, darker circle represents actors with more connections. For example, student -J- has the highest number of connections (9), which means that she/he is the most book-sharing person in the network. Connection frequency can be represented by the thickness (weighted) of the line showing the connection (Grandjean, 2021). The figure shows that Student -J- has a more frequent connection with Student -A- and Student -D-.

Figure 8.8. Simulation of Typhoon Mawar (2005)[10]

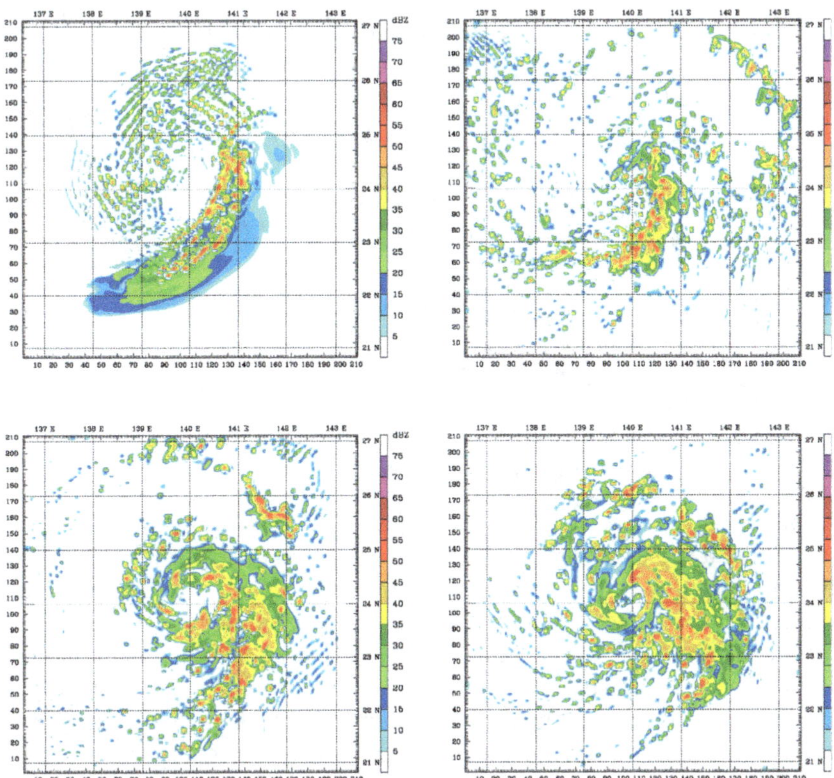

[10] Image source: Adapted from the original simulation image at Wikimedia. Licensed under the Creative Commons Attribution-Share Alike 3.0 license. To view a copy of this license, visit https://creativecommons.org/licenses/by-sa/3.0. Any changes made to the original material are noted, and this adapted work is distributed under the same Creative Commons Attribution-Share Alike 3.0 license.

Betweenness centrality refers to how often a node serves as a bridge between separate clusters within a network. Betweenness centrality differs from degree centrality. While a high degree of centrality in a network is associated with the number of connections a node has, betweenness centrality pertains to a node's role in connecting two distinct network clusters. A node may possess high betweenness centrality despite having a low degree centrality (Farahani, Karwowski, and Lighthall, 2019; Grandjean, 2021). Figure 8.7 shows that although actor -M- has a lower degree of centrality than actor -L- and actor -N-, it has more betweenness centrality.

There are many types of illustrations. Figure 8.8 exhibits a typhoon simulation. Figure 8.9 exhibits a vector illustration of tectonics.

Figure 8.9. A vector illustration of tectonics[11]

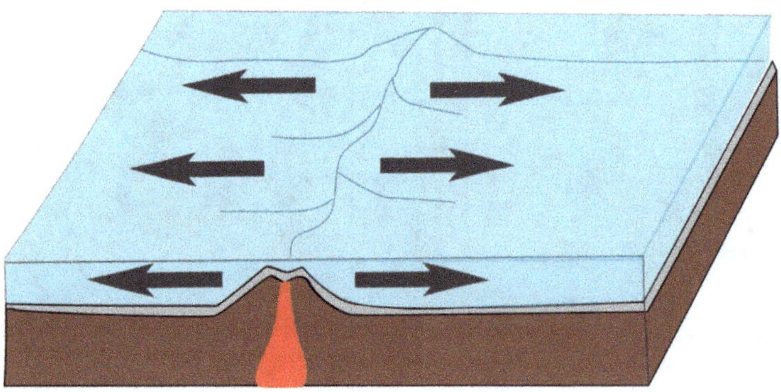

CONCLUDING REMARKS

Recognizing pseudoscience practices, being able to identify pseudoscientific content, and even recognizing the art of deception that surrounds our daily lives are important for both the defense of science and the defense of our minds. Because first, we must know the enemy, and the enemy (pseudoscience, art of deception, and ignorance) is everywhere. I pay special attention to the problem of information pollution, which I discussed in the preface section of the book. I see this as a potential threat to our general knowledge, which affects not only the pursuit of science but also our worldview. A scientist's mind should be less confused in certain respects. In other words, the ability to identify pseudoscience activities and misleading pieces of information must be developed. Doubt and verification from alternative sources should actually become a part of our daily information, beyond scientific activities. Because

[11] Image source: https://pixabay.com/vectors/dorsal-tectonics-plates-geology-6801801/

the misleading information that surrounds us not only affects our worldview, but also harms our healthy reasoning capability. People who cannot reason clearly cannot be expected to make healthy democratic choices. Scientists who cannot think clearly and form opinions based on incorrect information cannot be expected to get rid of their prejudices or biases. That's why we must put on our armor. Reasoning, doubt and verification are our armor, and if we recognize the enemy (pseudoscience and the art of deception), we are on the right track.

SUMMARY OF CHAPTER 8

Ignorance is something that most people see in others, not in themselves. Dunning-Kruger (2011) asserts that people with substantial deficits in their knowledge or expertise are not able to recognize those deficits. It is a big problem when a scientist is not able to recognize her/his deficits. Scientific studies require eliminating theoretical and methodological errors in the process of pursuing science. If a scientist has deficits related to her/his profession, this reflects in the work done by her/him. It may not be so easy to cope with individual deficits, but striving to overcome those deficits is important and the first step is to detect those deficits.

This book aims to protect the minds of young researchers from both the *art of deception* and *pseudoscience.* Firstly, there are many possible ways of the art of deception. The lie is the simplest form of deception. Manipulation, incomplete data, exaggeration, demagogy, propaganda, and ad hominem can be included as other forms of the art of deception. There are many tools to support deceptive statements. For example, visualization is one of those tools that can easily be used to mislead people. A researcher should be able to recognize the art of deception to avoid it.

Pseudoscience, or fake science, is very prevalent in the world outside of the professional science society, even in this modern age. Many people fall prey to pseudoscience. Those activities affect non-scientifically educated people. Even if most young researchers know their nature, it is important to warn them about their manipulative potential of pseudoscience. Researchers are urged to avoid activities contributing to fake science. Excluding activities of pseudoscience is an ethical standard for conducting accurate and valid science.

The importance of *visualizing information* can be seen as controversial. However, I believe that the presentation of processed data is one of the indicators of the quality of a study. Visualization is also an effective communication tool and can facilitate effective scientific communication. Numerous efficient computer and software technologies are available for powerfully visualizing information in different forms for researchers to benefit from.

FINAL NOTES

There are many scientific research and methodology books on the market. This diverse outline stems from the author's conceptualization of the scientific methods from her/his personal perspectives. Therefore, readers do not encounter a standard framework as in many methodology books. While some books provide a general idea about research methods in social sciences, some do present field-specific (e.g., medical, educational or business administration) approaches to scientific methods. This book presents a general look at research methods along with related concepts and topics to bring its readers to the state of scientific thought. Accordingly, I strongly recommend readers read field/ subject-specific methodology books after reading this introductory-level book. This book, as mentioned before, serves to prepare readers for a scientific mindset and provides basic concepts. Readers should keep in mind that each field and subject can necessitate unique approaches to conducting a research study. Research design, data collection method, analysis method, risks, and ethical principles can differ for different fields.

VOCABULARY FOR METHODOLOGY AND SCIENTIFIC THOUGHT

Absolute zero: The natural (real) zero point representing the absence of a feature. For example, the frequency or unit number could be the absolute zero point. The absence of certain types of individuals among the participants of a survey we conducted (such as Chinese, Black, or doctorate holder) is an example of the absolute zero point. (See also arbitrary zero.)

Ad hoc: Research that is conducted for a specific purpose to measure certain situations. In Latin, *ad hoc* means *for a specific reason.*

Ad hominem: This means *personal* and indicates an argument against a person who makes a claim (not against the claim itself).

Affective: This means *based on the senses.*

Analogy: This method of reasoning allows for systematic linking between a *source* and a *target.*

Anthology: A work created by selecting beautiful pieces from literary products and arranging them according to various methods.

Arbitrary zero: An artificially defined zero point that does not report the absence of a feature. For example, the zero point of a temperature scale does not indicate that the temperature is absent; it is an artificially defined point (based on the freezing point of water in the case of Celsius and Fahrenheit scales). (See also absolute zero.)

Argument: A sequence of statements in which a conclusion is inferred or supported by others (the premises). Arguments are pieces of reasoning.

Bias: In a study, bias is the influence that distorts the findings, leading them to deviate from the true or anticipated outcomes.

Bias error: A consistent measurement error that reduces the validity of a statistical test.

Bibliometric: The use of statistical methods to analyze books, articles, and other publications.

Causality: The relation of dependence and production between variables or events. In a causation relationship, a change in one variable (the cause) brings about a change in the other variable.

Ceteris paribus: This is Latin for *other things being equal.*

Coding: It refers to the process of categorizing and organizing data into meaningful groups or themes. This process involves systematically analyzing qualitative or quantitative data to identify patterns, themes, or relationships.

Concept: An idea or principle that is connected with something *abstract.*

Confirmatory research: This research is conducted to determine the relative accuracy of a theory by testing hypotheses derived from it.

Correlation: It indicates the relationship between two variables that are statistically dependent.

Deduction: This is a way of reaching a judgment about individual events by acting based on general principles. The expression *"All people are mortal. Socrates is a human, so Socrates is mortal"* is an example of deduction.

Dependent variable: The dependent variable, also known as the outcome variable, represents the response, behavior, or outcome under investigation in research. The assumption here is that the dependent variable is to be influenced or caused by variations in the independent variable. The dependent variable can be studied for explanation or prediction purposes.

Determinism: This is a philosophy of science that argues that natural events and human behavior result from a series of preliminary events/factors, and that these can be determined using basic scientific laws.

Dialectic: The art of systematic discussion. The dialectic was first defined by Aristotle. In terms of scientific research, dialectic is related to an inquiry with the aim *of finding a method to argue about a proposed problem from probable premises while avoiding self-contradiction* (Nicolaou 2011).

Dichotomy: This involves splitting something into two separate groups (or divisions) in scientific theories or approaches, such as positive or normative forms of scientific fields. For example, in biology, organisms can be classified into two broad categories based on their cellular structure: prokaryotes and eukaryotes. Prokaryotes are simple cells lacking a true nucleus and membrane-bound organelles such as bacteria and archaea. Eukaryotes are complex cells with a true nucleus and membrane-bound organelles such as plants and animals. This classification provides a dichotomy between two fundamentally different types of cells based on their internal structure. In psychology, personality traits are often categorized using a dichotomy known as the Big Five personality traits, also known as the Five-Factor Model. These traits are considered to be the fundamental dimensions of personality, such as (1) Openness to Experience vs. Closedness, (2) Conscientiousness vs. Lack of

Conscientiousness, (3) Extraversion vs. Introversion, (4) Agreeableness vs. Disagreeableness, (5) Neuroticism vs. Emotional Stability.

Double-blind: This is an application made for the reliability of the research, and is based on the fact that the research staff takes measurements without knowing who is in the experimental and control groups. For example, doctors who determine which diseases have been alleviated when testing a new drug should not know which patients have been given the drug.

Empirical research: This research is based on a direct collection of data (such as through a survey, observation, interviews, or document review).

Epistemology: This is the science of knowledge, and involves studying the definition of knowledge, its sources, and the standards of valid and reliable knowledge.

Ethics: Ethics examines how moral principles will operate (or be employed) under given conditions. For example, the expression of professional ethics refers to the determination of how general moral principles will be applied in the performance of a profession.

Etymology: This science studies the origin of words.

Experimental group: This is the group on which the effect of variables is tested. In other words, it is the group in which the relationship between the independent variable and the dependent variable is examined. For example, when a researcher tries to prove the hypothesis "*Fatigue prevents learning*" through an experiment, *fatigue* is the independent variable, and *learning* is the dependent variable. While the control group continues to learn via normal activities, the learning group is tested by increasing and/or decreasing fatigue. The point is that the average intelligence level should show a normal distribution in all groups.

Frequency: This shows the number of repetition units of qualitative (categorical) data.

Frequency distribution: This entails summarizing qualitative data using column or circle charts. If there are too many categories in a dataset, it is practical and widely preferred to combine those below a certain frequency value (usually below 5%) in another category. Examples that indicate frequency distribution include the following: automobile models sold by a dealer in 2014, types of errors frequently encountered in a production process, letter grades in an English course, and the wait times (in minutes) of customers using a company's customer service.

Full count: This is the collection of data from each member of a population regarding a particular subject (a feature of the population). The census is an example of a full count.

Goldilocks principle: This refers to the tendency of an observation to be close to central values rather than extreme ones.

Grounded theory: This is based on data; it is a qualitative research technique based on induction theory development. The researchers coined this name because they tried to extract a theory from embedded data.

Heterodox: This refers to a thought that has deviated from the mainstream. It consists of the Greek *heteros* meaning "different" and *doxa* meaning "teaching, thought."

Hypothesis: This is a proposition that needs to be verified or falsified to solve a research problem. A hypothesis is a specific estimation of the relationship between two or more variables based on a theory or previous research. A hypothesis emerges by formulating a thesis.

Induction: Founded by Francis Bacon, this is a way of reaching general propositions by acting on individual events. In other words, it is a way of reaching a statement about the whole or reaching a single statement, thereby achieving consistency in fiction, structure, and thought. The statement "*People need water, animals need water, and plants need water, so all living things need water*" is an example of induction.

Intervening variable: Such a variable affects the causal relationship between other variables. It is also known as a *mediator* or *intermediary* variable.

Laws (scientific laws): These are repetitive (and also universal) generalizations. The *Law of Gravity* is an example. Gravity is a law of physics that covers (i.e., is generalizable to) all objects.

Measurement: The process of converting observations into numbers based on a set of rules.

Measurement error: This error expresses the differences in scores obtained from the same individuals using the same measurement tool.

Meta-analysis: An analysis technique that combines findings of multiple quantitative studies.

Morality: Basic principles and rules that define what is right and what is wrong.

Nomothetic: This refers to the objectivist view of science that methodologically requires universal laws to be studied and formulated.

Occult: This means the search for *secret information* by means other than the scientific method. It is a false technique in science.

Ockham's razor (i.e., the principle of parsimony): This theory is founded upon on the philosophical position that *"In an environment where everything is equal, the simplest explanation is the most likely to be true."* Briefly, "Keep it simple!" It is preferable to use one explanation (versus two) to explain a phenomenon. In other words, if a scientific result or explanation is simple when it is based on relatively few assumptions, one should not refer to unobservable causes and only refer to specific reasons. For example, it is more acceptable to use the general laws of motion previously developed to explain the Earth's motion in space than to develop brand new assumptions specific to this situation.

Oxymoron: This is a statement combining one concept that cannot be logically juxtaposed with another concept. It is a rhetorical art form based on the combination of contradictions.

Paradigm: This is a set of concepts or thought patterns, including theories, research methods, postulates, and standards to contribute to a field. It is also expressed as the perspective or sum of common arguments on a particular topic or field.

Parameter: A numerical characteristic of a population or model (i.e., general inferences about a population). (Numerical) inferences that represent any property (variable) of the population are parameters.

Peer review: This is a process in which a scientific study (article or book) is examined by professionals in the field.

Philosophy: This is the field of study that examines the subjects of existence (metaphysics), ethics, aesthetics, epistemology, and politics based on logical principles and reasoning.

Philosophy of science: This is a branch of philosophy that examines the foundations, methods, and implications of science.

Polymath: This type of person specializes in multiple branches of a field.

Positivism: This approach reduces propositions to simple facts. Positivism is a philosophy of science based on making analyses and producing generalizable inferences.

Post hoc/ergo propter hoc fallacy: This is a form of logical inference. It simply means that *one event comes after another event, so the first event is the cause.* However, even if one event occurs before another, it is wrong to claim on that basis that it is the cause of the second event.

Premise: This is a scientific argument asserted as a pre-assumption.

Principle: This is the general basis or starting point that guides the researcher in both theoretical studies and application activities. Principles such as objectivity-impartiality and ethical principles in science arise from laws and theories, and lead to subsequent laws and theories.

Propositional logic: This form of logic proceeds in a bivalent way; propositions are either *true* or *false*.

Questionnaire: This is a structured set of questions directed at participants to collect data on a subject (such as the attitudes, thoughts and feelings, and suggestions of consumers, employees, or managers).

Reasoning: This is the process of reaching a conclusion from at least two thoughts, where one is treated as the proof of the other.

Reductio ad absurdum (proof by contradiction): This means reduction to the absent or absurd. It is a logical method in which the first claim is accepted as true, and a ridiculous result is reached based on the first claim. The first claim is then deemed false.

Reduction: A procedure whereby a given domain of items can be 'reduced' to the other. For example, numbers can be reduced to sets, and hence, number theory can be reduced to set theory.

Regression analysis: This analytical method refers to the tendency to the mean. In this regard, it indicates a natural distribution and is a symmetrical measurement.

Reliability: This refers to the consistency of the measurement instrument (test or scale), which is considered reliable if the measurement instrument produces the same results when the outcome is repeated under the same conditions. Therefore, a scale gains reliability only by producing the same outcome repeatedly under similar conditions. The correlation between the results of the measurements reveals reliability.

Replication: This is the repetition of a study with a new sample and sometimes under new conditions.

Research: This is the process of generating knowledge of facts, objects, and events.

Sample: A cluster that is randomly or systematically taken from a particular population. The better the selected sample represents the population, the more accurate the sample results represent the fact (the measured feature or variable) of the population.

Sampling: The process of selecting some part of a population to estimate studied subject about the population. It is very important for a research study to appoint a sample to represent the population most accurately. Sampling can be conducted in several ways, such as random sampling or cluster sampling. **Random sampling** is based on the selection of people randomly from a population instead of creating a sample based on certain parameters. **Cluster sampling** is a method of sampling in which the members of a population constitute a cluster naturally. Families or members of a particular school can be given as examples for clusters. After a number of clusters are selected randomly then the chosen ones constitute a subsample.

Semantics: This is a field of science and philosophy that studies the meaning of words.

Survey study/research: This is a type of research study based on collecting data through a questionnaire.

Syllogism: This is a deductive form of reasoning to reach a conclusion and to draw inferences. A syllogism involves drawing a third proposition, or conclusion, from two distinct but related premises. Valid propositions involve predicates and quantifiers (Antonelli 2008).

Taxonomy/typology/topology: **Taxonomy** is the science of classification, which studies the subspecies of a particular species. It is used in both the natural and social sciences. It is recognized as a branch of science used to classify plants and animals, especially in biology in the natural sciences. Taxonomy is also used in the natural sciences other than livings. Rock types in geology and star types in astronomy are examples of taxonomy. It is also used in the social sciences in classifying languages in the field of philology, and in classifying certain approaches (schools) in economics and sociology. **Typology** focuses on object/system/process categorization (classification). The typological distinctions are qualitative (living and inanimate, such as plants and animals) and aim to reveal categories (such as birds, fish, etc.) instead of grading, as in topology. As a result, it defines discrete variants. **Topology** refers to a degree of distinction, rather than categorizing species according to their contrasts, for example, color density, texture and shape, or relative values in mathematics, such as in drawing or painting. In this regard, topological parameters are based on continuous variability (like shades of a color scale); for this reason, it is more correct to express a given range or feature in degrees. With this feature, topological studies focus on the quantitative differences of the object/system/process studied. Other indicators of topological studies are trends and continuity in transformation. There are also sequences in taxonomy. The order in which the fundaments are considered is important. This is a feature that distinguishes typology from the taxonomy. In short, typology classifies types according to their qualities (e.g., warm vs. cool colors), taxonomy again classifies them according to their qualities (e.g., ordering colors from warm to cold, or ordering colors according to their wavelength in physics), and finally, topology creates ratings (such as shades of yellow from lighter to darker along a gradient) that are quantitative in this respect.

Testimonial evidence: An explanation about the reality of a fact or claim based on one's personal experience.

Theory: This is a knowledge network developed to explain the relationships between the variables that are basic to science in the most simple and understandable way. Cause-effect relationships between events are explained with the help of theories. A theory is a thought system developed based on previous observations and experiments to perceive, explain, and define events. A theory is general, rational, and abstract (since it is included in the systematic nature of thought). A theory is expressed as a hypothesis if it is not yet proven by research.

Theme: The subject or main idea in a talk, piece of writing, or work of art.

Thesis: This is the proof of an idea, a thought, or a subject by defending it based on scientific evidence or logical inferences.

Validity: This refers to the suitability of a scale (or test) used to measure what is intended to be measured.

Variance: This is a measure of the frequency distribution of the scores obtained. It is based on determining the deviation from the general mean by calculating the sum of the squares of the distance of each score from the average. It is a measure of the variability of population measurements.

BIBLIOGRAPHY

Abrams, L., & McMillian J. (2015). Ethical issues, principles, and practices, In *Fundamentals of educational research*. Boston: Pearson.

Ajmani, V. (2009). *Applied econometrics using the SAS system*. John Wiley & Sons.

Alberts B, Johnson A, Lewis J, et al. (2002). *Molecular biology of the cell*. New York: Garland Science.

Allchin, D. (2012). Teaching the nature of science through scientific errors. *Science Education*, 96(5), 904-926.

Allum, N. (2011). What makes some people think astrology is scientific?. *Science Communication*, 33(3), 341-366.

Alston, W. P. (2005). Empiricism. In E. Craig (Ed.), *The Shorter Routledge Encyclopedia of Philosophy*, London: Routledge.

Alvesson, M., & Deetz, S. (2020). *Doing critical research*, SAGE Publications Limited.

American Psychological Association. (2020). *Publication manual of the American Psychological Association* (7th ed.). APA.

Amijee, F. (2021). Explaining contingent facts. *Philosophical Studies*, 178(4), 1163-1181.

Anderson, M. (2007). *Quantitative history*. In W. Outhwaite & S. Turner (Eds.). *The Sage Handbook of Social Science Methodology*. 246-263, London: Sage.

Ankeny, R. A., & Leonelli, S. (2016). Repertoires: A post-Kuhnian perspective on scientific change and collaborative research. *Studies in History and Philosophy of Science Part A*, 60, 18-28.

Arfini, S., & Magnani, L. (2020). Introduction: knowing the unknown: Philosophical perspectives on ignorance. *Synthese*, 199, 689-693.

Asimov, I. (1988). *Isaac Asimov's Book of Science and Nature Quotations*. p 281, Weidenfeld & Nicolson, New York.

Aspers, P., & Corte, U. (2019). What is qualitative in qualitative research. *Qualitative Sociology*, 42(2), 139-160.

Åsvoll, H. (2014). Abduction, deduction and induction: can these concepts be used for an understanding of methodological processes in interpretative case studies?. *International Journal of Qualitative Studies in Education*, 27(3), 289-307.

Aubusson, P. J., Aubusson, P., Harrison, A. G., & Ritchie, S. M. (Eds.). (2006). *Metaphor and analogy in science education* (Vol. 30). Springer Science & Business Media.

Audi, R. (2010). Epistemology: A contemporary introduction to the theory of knowledge. Routledge.

Azzam, T., Evergreen, S., Germuth, A. A., & Kistler, S. J. (2013). Data visualization and evaluation. *New Directions for Evaluation*, 2013(139), 7-32.

Bach, K. (2006). Language, logic, and form. *A companion to philosophical logic*, 49-72.

Barker, G. & Kitcher, P. (2014). *Philosophy of Science: A New Introduction.* Oxford University Press.

Bailer-Jones, D. M. (2003). When scientific models represent. *International Studies in the Philosophy of Science,* 17(1), 59-74.

Balas, E. A. (2018). *Innovative Research in Life Sciences: Pathways to Scientific Impact, Public Health Improvement, and Economic Progress.* John Wiley & Sons.

Ball, L. (1993). The dynamics of high inflation (No. w4578). *National Bureau of Economic Research.*

Barton, J., & Haslett, T. (2007). Analysis, synthesis, systems thinking and the scientific method: rediscovering the importance of open systems. Systems Research and Behavioral Science: *The Official Journal of the International Federation for Systems Research,* 24(2), 143-155.

Basile, G., & Marro, G. (1992). *Controlled and conditioned invariants in linear system theory.* New Jersey: Prentice Hall.

Baur, N., (2011). Variables. In George Ritzer, J. Michael Ryan (Eds.), *The Concise Encyclopedia of Sociology,* 676-677, John Wiley & Sons.

Bayne, T., Brainard, D., Byrne, R. W., Chittka, L., Clayton, N., Heyes, C., ... & Webb, B. (2019). What is cognition?. *Current Biology,* 29(13), R608-R615.

Beach, D. (1996). *The responsible conduct of research.* Weinheim, New York.

Bechtel, W., & Hamilton, A. (2007). Reduction, integration, and the unity of science: Natural, behavioral, and social sciences and the humanities. Kuipers, Theo. A. F. (Ed.), *General Philosophy of Science: Focal Issues.* 377-430. Elsevier.

Bennett, B. (2012). *Logically fallacious: the ultimate collection of over 300 logical fallacies,* Academic Edition.

Bermudez, J. L. (2008). Cartesian Skepticism: arguments and antecedents. *The Oxford Handbook of Skepticism,* 53-79.

Bickman, L., & Rog, D. J. (2008). *The SAGE handbook of applied social research methods,* (2nd ed.), Sage Publications.

Bishop, M., & Trout, J. D. (2016). Epistemology for (Real) People. *A Companion to Applied Philosophy,* 103-119.

Blaug, M. (1992). *The methodology of economics: Or, how economists explain.* Cambridge University Press.

Bogdan, R., & Biklen, S. K. (1997). *Qualitative research for education.* Boston, MA: Allyn & Bacon.

Bohm, D. (1961). On the relationship between methodology in scientific research and the content of scientific knowledge. *The British Journal for the Philosophy of Science,* 12(46), 103-116.

Bokulich, A. (2011). How scientific models can explain. *Synthese,* 180(1), 33-45.

Bourdeau, M. (2022). Auguste Comte. In E. N. Zalta, (Ed.), *The Stanford Encyclopedia of Philosophy,* Retrieved 17 September 2023, from https://plato.stanford.edu/entries/comte/

Bowerman, B. L., O'Connell, R. T., Murpree, E. S., & Orris, J. B. (2004). *Essentials of Business Statistics.* McGraw-Hill.

Brigandt, I. & Love, A. (2023). Reductionism in Biology, In Zalta, E. N. & Nodelman, U. (Eds.), *The Stanford Encyclopedia of Philosophy,* Retrieved January

21, 2024, from https://plato.stanford.edu/archives/sum2023/entries/reduction-biology/>

Britannica, (n.d.). Intension and extension. In *Britannica Encyclopedia.* Retrieved October 3, 2020, from https://www.britannica.com/topic/intension

Britannica, (n.d.). Ontology. In *Britannica Encyclopedia.* Retrieved March 7, 2018, from https://www.britannica.com/topic/ontology-metaphysics

Britannica, (n.d.). Taxonomy. In *Britannica Encyclopedia.* Retrieved June 27, 2021, from https://www.britannica.com/science/taxonomy

Britannica, (n.d.). Technology. In *Britannica Encyclopedia.* Retrieved April 8, 2019, from https://www.britannica.com/technology/technology

Brooklyn, (n.d.). Topology & taxonomy. In *Brooklyn Academic.* Retrieved July 20, 2019, from http://academic.brooklyn.cuny.edu/education/jlemke/webs/wess/tsld002.htm

Brooks, H. (1994). The relationship between science and technology. *Research Policy,* 23(5), 477-486.

Bryant, R. E. (1997). *Discovery and decision: exploring the metaphysics and epistemology of scientific classification,* Doctoral dissertation, University of Edinburgh.

Bryman, A. (2007). The research question in social research: what is its role?. *International Journal of Social Research Methodology,* 10(1), 5-20.

Buhringer, G., & Sassen, M. (2010). Reliability and validity. Miller, P. G., Strang, J., & Miller, P. M. (Eds.). *Addiction Research Methods.* John Wiley & Sons.

Buller, D. B., & Burgoon, J. K. (1996). Interpersonal deception theory. *Communication Theory,* 6(3), 203-242.

Bunge, M. (1966). Technology as applied science. In *Contributions to a Philosophy of Technology,* 19-39, Springer.

Butz, M. R. (1995). Chaos theory, philosophically old, scientifically new. *Counseling and Values,* 39(2), 84-98.

Cairo, A. (2016). *The truthful art: Data, charts, and maps for communication.* New Riders.

Callender, C. (2011). Philosophy of science and metaphysics. *The Continuum Companion to the Philosophy of Science,* 33-54.

Cambridge Dictionary (n.d.). Analytical, In *Cambridge Dictionary.* Retrieved October 19, 2023, from https://dictionary.cambridge.org/us/dictionary/english/analytical

Cambridge Dictionary (n.d.). Lateral thinking. In *Cambridge Dictionary.* Retrieved July 17, 2020, from https://dictionary.cambridge.org/us/dictionary/english/lateral-thinking

Cambridge Dictionary (n.d.). Philosophy. In *Cambridge Dictionary.* Retrieved May 19, 2019, from https://dictionary.cambridge.org/us/dictionary/english/philosophy

Cambridge Dictionary (n.d.). Thinking. In *Cambridge Dictionary.* Retrieved July 17, 2020, from https://dictionary.cambridge.org/us/dictionary/english/thinking?q=Thinking+

Carl Sagan Quote. (1979). In C. C. Gaither, & A. E. Cavazos-Gaither (Eds.), *The Gaither's dictionary of scientific quotations.* (2008). p 451. Springer.

Cash, P. (2020). Where next for design research? Understanding research impact and theory building. *Design Studies*, 68, 113-141.

Castellan, C. M. (2010). Quantitative and qualitative research: A view for clarity. *International Journal of Education*, 2(2), 1-14.

Chakravartty, A. (2017). Scientific Realism, In E. N. Zalta, (Ed.), *The Stanford Encyclopedia of Philosophy*, Retrieved March 19, 2019, from https://plato.stanford.edu/archives/sum2017/entries/scientific-realism/

Chen, D., & Stroup, W. (1993). General system theory: Toward a conceptual framework for science and technology education for all. *Journal of Science Education and Technology*, 2(3), 447-459.

Chigbu, U. E. (2019). Visually hypothesizing in scientific paper writing: Confirming and refuting qualitative research hypotheses using diagrams. *Publications*, 7(22).

Cohen, L. Manion. L., & Morrison, K. (2011). *Research Methods in Education.* London: Routledge.

Cohen, S. M. (2004). *Glossary*, https://faculty.washington.edu/smcohen/120/Glossary.pdf

Collier, D. Laporte, J., & Seawright, J., (2008). Typologies: Forming concepts and creating categorical variables. In Janet M. Box-Steffensmeier, Henry E. Brady, David Collier, (Eds.), *Oxford Handbooks of Political Methodology*, Oxford: Oxford University Press.

Connolly, P. *John Locke* (1632—1704), https://iep.utm.edu/locke

Correll, M., & Heer, J. (2017). Black hat visualization. In *Workshop on Dealing with Cognitive Biases in Visualizations*, 1 (3).

Cowling, S. (2017). *Abstract entities.* New York: Routledge.

Coyne, J. A. (2015). *Faith Versus Fact: Why Science and Religion Are Incompatible.* Penguin Books.

Craig, E. (2005). Ontology. In E. Craig (Ed.), *The shorter Routledge Encyclopedia of Philosophy*, London: Routledge.

Creath, R. (2023). Logical Empiricism, In E. N. Zalta & U. Nodelman (Eds.), *The Stanford Encyclopedia of Philosophy*, Retrieved November 22, 2023, from https://plato.stanford.edu/archives/win2023/entries/logical-empiricism/

Creswell, J. W. (2007). Five qualitative approaches to inquiry. Qualitative inquiry and research design: Choosing among five approaches, 2, 53-80.

Creswell, J. W., & Creswell, J. D. (2017). Research design: *Qualitative, quantitative, and mixed methods approaches.* Sage Publications.

Crimmins, M. (2005). Semantics. In E. Craig (Ed.), *The shorter Routledge Encyclopedia of Philosophy*, London: Routledge.

Crupi, V., Fitelson, B., & Tentori, K. (2008). Probability, confirmation, and the conjunction fallacy. *Thinking & Reasoning*, 14(2), 182-199.

Cruz, J. (2006). Epistemology, In L. Nadel (Ed.), *Encyclopedia of Cognitive Science*, 1-6, New York: Nature Publication Group.

Cummings, L. (1998). The scientific reductionism of relevance theory: The lesson from logical positivism. *Journal of Pragmatics*, 29(1), 1-12.

Cunningham, C. J., Weathington, B. L., & Pittenger, D. J. (2013). *Understanding and conducting research in the health sciences.* John Wiley & Sons.

Da Silva, S. L. C., & Amaral, F. G. (2019). Critical factors of success and barriers to the implementation of occupational health and safety management systems: A systematic review of literature. *Safety Science*, 117, 123-132.

Dahlberg, K. (1995). Qualitative methodology as caring science methodology. *Scandinavian Journal of Caring Sciences*, 9(3), 187-191.

Davidson, R. (2014). Using infographics in the science classroom. *The Science Teacher*, 81(3), 34.

Day, R. H. (1993). The Ames room from another viewpoint. *Perception*, (22), 1007-1011.

De Gruijter, D. N., & Kamp, L., J. T. (2007). *Statistical test theory for the behavioral sciences*. New York: Chapman and Hall/CRC.

De Pierris, G. (2006). Hume and Locke on scientific methodology: The Newtonian legacy, *Hume Studies*, 32(2), 277-329.

Deery, O., & Nahmias, E. (2017). Defeating manipulation arguments: Interventionist causation and compatibilist sourcehood, *Philosophical Studies*, 174(5), 1255-1276.

Dembski, W. A. (2005). Randomness. In E. Craig (Ed.), *The shorter Routledge Encyclopedia of Philosophy*, London: Routledge.

DeWitt, R. (2010). Philosophy of science. In F. Allhoff (Ed.), *Philosophies of the Sciences. A Guide*, Wiley.

D'ignazio, C. (2019). Data visualization. In Hobbs, R. & Mihailidis, P. (Eds.), Gianna Cappello, Maria Ranieri, and Benjamin Thevenin (Assoc Eds.), *The International Encyclopedia of Media Literacy*, John Wiley & Sons.

Dornbusch, R., & Reynoso, A. (1989). Financial factors in economic development. *The American Economic Review*, 79(2), 204-209.

Drack, M., & Schwarz, G. (2010). Recent developments in general system theory. *Systems Research and Behavioral Science*, 27(6), 601-610.

Dubé, J. P. (2004). Multiple discreteness and product differentiation: Demand for carbonated soft drinks. *Marketing Science*, 23(1), 66-81.

Dunning, T., & Freedman, D. A. (2007). Modeling selection effects. In W. Outhwaite & S. Turner (Eds.), *The Sage Handbook of Social Science Methodology*. London: Sage.

Dunning, D. (2011). *The Dunning–Kruger effect: On being ignorant of one's own ignorance. In Advances in experimental social psychology*, 44, 247-296, Academic Press.

Easton, G. (2010). Critical realism in case study research. *Industrial Marketing Management*, 39(1), 118-128.

Ekman, P. (2009). *Telling lies: Clues to deceit in the marketplace, politics, and marriage*. New York: Norton.

Emanuel, K. (2013). Tropical cyclones. *Annual Review of Earth and Planetary Sciences*, 31(1), 75-104.

Emanuel, K., Ravela, S., Vivant, E., & Risi, C. (2006). A statistical deterministic approach to hurricane risk assessment. Bulletin of the American Meteorological Society, 87(3), 299-314.

Evans, J. S. B. (2005). Deductive reasoning. In E. Craig (Ed.), *The Shorter Routledge Encyclopedia of Philosophy*, London: Routledge.

Evans, J. S. B., & Over, D. E. (2013). Reasoning to and from belief: Deduction and induction are still distinct. *Thinking & Reasoning*, 19(3-4), 267-283.

Evergreen, S., & Metzner, C. (2013). Design principles for data visualization in evaluation. *New Directions for Evaluation*, 2013(140), 5-20.

Everitt, B., & Skrondal, A. (2002). *The Cambridge Dictionary of Statistics*, Cambridge: Cambridge University Press.

Ewoldsen, D. R. (2017). Normal science and paradigm shift. *The International Encyclopedia of Communication Research Methods*, 1-17.

Exner, R. M., & Rosskopf, M. F. (2011). *Logic in elementary mathematics*. Courier Corporation.

Fantl, J., & McGrath, M. (2002). Evidence, pragmatics, and justification. *The Philosophical Review*, 111(1), 67-94.

Farahani, F. V., Karwowski, W., & Lighthall, N. R. (2019). Application of graph theory for identifying connectivity patterns in human brain networks: a systematic review. *Frontiers in Neuroscience*, 13.

Fine, A. (2005). Scientific realism and antirealism. In E. Craig (Ed.). *The Shorter Routledge Encyclopedia of Philosophy*, London: Routledge.

Fine, K. (2011). What is metaphysics?. In Tahko, T. (Ed.), *Contemporary Aristotelian Metaphysics*, 8-25.

Fox, N. J. (2008). Post-positivism, Lisa M. Given (Ed.), *The SAGE Encyclopedia of Qualitative Research Methods*, 659-664, California: Sage Publications.

Fraenkel, J. R., Wallen, N. E., & Hyun, H. H. (2011). *How to design and evaluate research in education*. New York: McGraw-Hill Humanities/Social Sciences/ Languages.

Franconeri, S. L., Padilla, L. M., Shah, P., Zacks, J. M., & Hullman, J. (2021). The science of visual data communication: What works. *Psychological Science in the Public Interest*, 22(3), 110-161.

Freedman, D., Pisani, R., & Purves, R. (1980). *Statistics*. New York: Norton.

Galbacs, P. (2020). *The Friedman-Lucas Transition in Macroeconomics: A Structuralist Approach*. Academic Press.

Galich, G., & Shnyakina, N. (2015). Categories of immaterial objects in mind and language. *Procedia-Social and Behavioral Sciences*, 206, 30-35.

Galileo Galilei Quote. (n.d.). In *Oxford Dictionary of Scientific Quotations*. W. F. Bynum and Roy Porter (Eds.). (2005). Oxford University Press.

Gay, L. R., Mills, G. E., & Airasian, P. W. (2011). *Educational research: competencies for analysis and applications*, Boston: Pearson.

Glänzel, W., & Schubert, A. (2003). A new classification scheme of science fields and subfields designed for scientometric evaluation purposes. *Scientometrics*, 56(3), 357-367.

Glesne, C. (2011). *Becoming Qualitative Researchers*, Boston: Pearson.

Golbeck, J. (2013). *Analyzing the social web*. Elsevier.

Goldberg, L. R. (1992). The development of markers for the Big-Five factor structure. *Psychological Assessment*, 4(1), 26.

Gooding, D. C. (2005). Thought experiments. In E. Craig (Ed.), *The Shorter Routledge Encyclopedia of Philosophy*, London: Routledge.

Gouinlock, J. (2005). Determinism and indeterminism. In E. Craig (Ed.), *The Shorter Routledge Encyclopedia of Philosophy*, London: Routledge.

Graham, G. (2014). Aesthetics as a Normative Science. *Royal Institute of Philosophy Supplements*, 75, 249-264.

Grandjean, M. (2021). Introduction to Social Network Analysis: Basics and Historical Specificities. *HNR+ResHist Conference 2021*, Historical Network Research, Luxembourg.

Grayling, A. C. (2002). Epistemology. In Nicholas Bunnin, E. P. Tsui-James (Eds.), *The Blackwell Companion to Philosophy*, 35-60, Blackwell Publishing.

Grene, M. (1999). Descartes and Skepticism. *The Review of Metaphysics*, 52(3), 553–571.

Grove, S. K., & Gray, J. R. (2018). *Understanding nursing research: Building an evidence-based practice*. E-book, Elsevier.

Haig, B.D. (2018). *The Philosophy of Quantitative Methods*. The Oxford handbook of quantitative methods, 7-30.

Hall, F. S., Drgonova, J., Jain, S., & Uhl, G. R. (2013). Implications of genome wide association studies for addiction: are our a priori assumptions all wrong?. *Pharmacology & Therapeutics*, 140(3), 267-279.

Hands, D. W. (2012). The positive-normative dichotomy and economics. *Handbook of the Philosophy of Science*, 13, 219-239.

Hannan, M. T., & Freeman, J. (1977). The population ecology of organizations. *American Journal of Sociology*, 82(5), 929-964.

Hayduk, L., & Pazderka-Robinson H. (2007). Fighting to understand the world causally: Three battles connected to the causal implications of Structural Equation Models. In Outhwaite, W., & Turner, S. (Eds.), *The Sage Handbook of Social Science Methodology*. London: Sage.

Henderson, D. (2007). Rationality and rationalist approaches in the social sciences. In W. Outhwaite & S. Turner (Eds.), *The Sage Handbook of Social Science Methodology*. London: Sage.

Herbert Spencer Quote. (1860). In C. C. Gaither, & A. E. Cavazos-Gaither (Eds.), *The Gaither's dictionary of scientific quotations*. (2008). p 733. Springer.

Higgins, J. P. (2002). Nonlinear systems in medicine. *The Yale Journal of Biology and Medicine*, 75, 247-260.

Hitchcock, D. (2024). Critical thinking. In E. N. Zalta & U. Nodelman (Eds.), *The Stanford Encyclopedia of Philosophy*, Retrieved June 21, 2024, from https://plato.stanford.edu/archives/sum2024/entries/critical-thinking

Hjørland, B., & Nicolaisen, J. (2004). Scientific and Scholarly Classifications are not "Naïve": a Comment to Begthol (2003). *Knowledge Organization*, 31 (1), 55-61.

Holliday, A. (2012). Validity in qualitative research. In Chapelle, C. A. (Ed), *The Encyclopedia of Applied Linguistics*, Wiley-Blackwell.

Holocaust Encyclopedia. (1947). Nuremberg codes. In *Holocaust Encyclopedia*. Retrieved August 5, 2020, from https://encyclopedia.ushmm.org/content/en/article/the-nuremberg-code

Holyoak, K. J. (2012). Analogy and relational reasoning. *The Oxford Handbook of Thinking and Reasoning*, 234-259.

Hon, G. (1989). Towards a typology of experimental errors: An epistemological view. *Studies in History and Philosophy of Science Part A,* 20(4), 469-504.

Hopf, H., Krief, A., Mehta, G., & Matlin, S. A. (2019). Fake science and the knowledge crisis: ignorance can be fatal. *Royal Society Open Science,* 6(5), 190161.

Høyrup, J. (1999). Pythagorean rule and" theorem: mirror of the relation between Babylonian and Greek Mathematics. In *Babylon: Focus mesopotamischer Geschichte, Wiege früher Gelehrsamkeit, Mythos in der Moderne: 2.* Internationales Colloquium der Deutschen Orient-Gesellschaft 24.-26. März 1998 in Berlin. SDV Saarländische Druckerei und Verlag GmbH.

Huby, P. M. (2004). Elementary logic in the Ancient World. *Bulletin of the Institute of Classical Studies,* 47(1), 119-128.

Hull, D. L. (2005). Taxonomy. In E. Craig (Ed.), *The Shorter Routledge Encyclopedia of Philosophy,* London: Routledge.

Illari, P., & Russo, F. (2014). *Causality: Philosophical theory meets scientific practice.* Oxford University Press, UK.

Jago, M. (2007). Formal logic. *Philosophy Insights,* Humanities-Ebooks.

Jevons, W. S. (1893). *Elementary Lessons in Logic,* McMillan, New York.

Jevons, W. S. (1958). Principles of Science. *Daedalus,* 87(4), 148-154.

Jones, C. (1990). *Probabilistic non-determinism.* PhD Thesis. University of Edinburgh.

Kahn, K. B. (2018). Understanding innovation. *Business Horizons,* 61(3), 453-460.

Kamakura, W. A. (2010). Common methods bias. Jagdish Sheth, Naresh Malhotra (Eds.). *Wiley international encyclopedia of marketing.* John Wiley & Sons.

Kaplan, S. (2016). Mixing quantitative and qualitative research. *Handbook of qualitative organizational research: Innovative pathways and methods,* 423-433.

Kaufman, A., Wood, C. H., & Theyel, G. (2000). Collaboration and technology linkages: a strategic supplier typology. *Strategic Management Journal,* 21(6), 649-663.

Katz, M. J. (2009). *From research to manuscript: A guide to scientific writing.* Springer Science & Business Media.

Kaushik, V., & Walsh, C. A. (2019). Pragmatism as a research paradigm and its implications for social work research. *Social Sciences,* 8(9), 255.

Kennedy, H., Hill, R. L., Aiello, G., & Allen, W. (2016). The work that visualisation conventions do. *Information, Communication & Society,* 19(6), 715-735.

Kiel, L. D., & Elliott, E. (1997). *Exploring nonlinear dynamics with a spreadsheet: A graphical view of chaos for beginners,* 19-29. Ann Arbor, MI: University of Michigan Press.

Kim, J. (2005). Reduction, problems of. In E. Craig (Ed.), *The Shorter Routledge Encyclopedia of Philosophy,* London: Routledge.

Kincaid, H. (2005). Positivism in social science. In E. Craig (Ed.), *The Shorter Routledge Encyclopedia of Philosophy,* London: Routledge.

Klein, P. (2005). Epistemology. In E. Craig (Ed.), *The Shorter Routledge Encyclopedia of Philosophy,* London: Routledge.

Koch, C. (2018). What is consciousness. *Nature,* 557(7704), 8-12.

Kochan, J. (2013). Subjectivity and emotion in scientific research. *Studies in History and Philosophy of Science Part A,* 44(3), 354-362.

Kogler, H. (2007). Understanding and interpretation. In W. Outhwaite & S. Turner (Eds.), *The Sage Handbook of Social Science Methodology.* London: Sage.

Kruger, J., & Dunning, D. (1999). Unskilled and unaware of it: how difficulties in recognizing one's own incompetence lead to inflated self-assessments. *Journal of Personality and Social Psychology,* 77(6), 1121.

Kundu, S. (2015). *An introduction to business statistics.* Course document. http://www.ddegjust.ac.in/studymaterial/mcom/mc-106.pdf

Kuhn, T. S. (1962). *The structure of scientific revolutions.* The University of Chicago Press.

Ladyman, J. (2007). Ontological, epistemological, and methodological positions. Kuipers, In Theo. A. F. (Ed.), *General Philosophy of Science: Focal Issues.* 303-376. Elsevier.

Lawson, A. E. (2005). What is the role of induction and deduction in reasoning and scientific inquiry?. *Journal of Research in Science Teaching,* 42(6), 716-740.

LeCun, Y., Bengio, Y., & Hinton, G. (2015). Deep learning. *Nature,* 521(7553), 436-444.

Lemon, J., Degenhardt, L., Slade, T., and Mills K. (2010). Quantitative data analysis. In Miller, P. G., Strang, J., & Miller, P. M. (Eds.), *Addiction Research Methods.* John Wiley & Sons.

Lewis-Beck, M., Bryman, A. E., & Liao, T. F. (2004). *The Sage Encyclopedia of Social Science Research Methods.* Sage Publications.

Lewis, C. I., Langford, C. H., & Lamprecht, P. (1959). *Symbolic Logic.* New York: Dover Publications.

Lipton, P. (2017). Inference to the best explanation. In W. H. Newton-Smith (Ed.), *A Companion to the Philosophy of Science,* 184-193.

List, C. (2014). Free will, determinism, and the possibility of doing otherwise. *Noûs,* 48(1), 156-178.

Logan, D. C. (2009). Known knowns, known unknowns, unknown unknowns and the propagation of scientific enquiry. *Journal of Experimental Botany,* 60(3), 712-714.

Lombrozo, T. (2012). Explanation and abductive inference. *The Oxford Handbook of Thinking and Reasoning.* Oxford University Press.

Look, B. (2020). Gottfried Wilhelm Leibniz, In E. N. Zalta, (Ed.), *The Stanford Encyclopedia of Philosophy,* Retrieved May 4, 2021, from https://plato.stanford.edu/entries/leibniz

Lormand, E. (2005). Consciousness. In E. Craig (Ed.), *The Shorter Routledge Encyclopedia of Philosophy,* London: Routledge.

Loux, M. J., & Crisp, T. M. (2017). *Metaphysics: A contemporary introduction.* Routledge.

Lu-Adler, H. (2021). Kant and the Principle of Sufficient Reason. *The Review of Metaphysics,* 74(3), 301-330.

Lumpkin, B. (1980). The Egyptians and Pythagorean triples. *Historia Mathematica,* 7(2), 186-187.

Lunde, P. (2009). *The Book of Codes: Understanding the World of Hidden Messages: an Illustrated Guide to Signs, Symbols, Ciphers, and Secret Languages.* Berkeley: University of California Press.

Lynch, M. (2007). Discourse Analysis. In W. Outhwaite& S. Turner (Eds.), *The Sage Handbook of Social Science Methodology.* London: Sage.

MacBride, F. (2005). The particular–universal distinction: A dogma of metaphysics?. *Mind,* 114(455), 565-614.

Mackie, P. (2005). Existence. In E. Craig (Ed.), *The Shorter Routledge Encyclopedia of Philosophy,* London: Routledge.

Mahoney, J., & Acosta, L. (2022). A regularity theory of causality for the social sciences. *Quality & Quantity,* 56(4), 1889-1911.

Marie Curie Quote. (1937). In C. C. Gaither, & A. E. Cavazos-Gaither (Eds.), *The Gaither's dictionary of scientific quotations.* (2008). p 157. Springer.

Marin, V. H. (1997). General system theory and the ecosystem concept. *Bulletin of the Ecological Society of America,* 78(1), 102-104.

Markie, P. (2005). Rationalism. In E. Craig (Ed.), *The Shorter Routledge Encyclopedia of Philosophy,* London: Routledge.

Marradi, A. (1990). Classification, typology, taxonomy. *Quality and Quantity,* 24(2), 129-157.

Marshall, S. (2012). Science, pseudo-science and urban design. *Urban Design International,* 17(4), 257-271.

Martinich, A. P. (2005). Metaphor. In E. Craig (Ed.), *The shorter Routledge Encyclopedia of Philosophy,* London: Routledge.

Marvin, S. (Ed.). (2011). *Dictionary of Scientific Principles.* John Wiley & Sons.

Masic, I. (2020). Plagiarism and how to avoid it. In Shoja, M. M., Arynchyna, A., Loukas, M., D'Antoni, A. V., Buerger, S. M., Karl, M., & Tubbs, R. S. (Eds.). *A Guide to the Scientific Career: Virtues, Communication, Research, and Academic Writing.* 163-177, John Wiley & Sons.

Matuszek, K. (2015). Ontology, reality and construction in Niklas Luhmann's theory. *Constructivist Foundations,* 10(2), 203-210.

Max Planck Quote. (1950). In *Oxford Dictionary of Scientific Quotations.* W. F. Bynum and Roy Porter (Eds.). (2005). Oxford University Press.

Maxwell, J. (1992). Understanding and validity in qualitative research. *Harvard Educational Review,* 62(3), 279-301.

Maxwell, J. A. (2013). *Qualitative research design: An interactive approach,* (Vol. 41), Sage Publications.

Maxwell, N. (2012). In praise of natural philosophy: A revolution for thought and life. *Philosophia,* 40, 705-715.

McCain, K. (2019). How do explanations lead to scientific knowledge? In K. McCain & K. Kampourakis (Eds.). *What is Scientific Knowledge?: An Introduction to Contemporary Epistemology of Science.* Routledge.

McCusker, K., & Gunaydin, S. (2015). Research using qualitative, quantitative or mixed methods and choice based on the research. *Perfusion,* 30(7), 537-542.

McKnight, D. H. (2011). Good science, bad science: Preventing paradigm paralysis and method-bias malaise. *International Journal of Accounting Information Systems,* 12(2), 84-89.

McMillan, J. H. (1999). *Unit of Analysis in Field Experiments: Some Design Considerations for Educational Researchers.*

McMillan, J. (2015). *Fundamentals of educational research.* Boston: Pearson.

McOmber, J. B. (1999). Technological autonomy and three definitions of technology. *Journal of Communication,* 49(3), 137-153.

Medin, D. L., & Rips, L. J. (2005). Concepts and categories: Memory, meaning, and metaphysics. *The Cambridge Handbook of Thinking and Reasoning,* 37-72.

Meeker, K. (2011). Quine on Hume and the analytic/synthetic distinction. *Philosophia,* 39(2), 369-373.

Mellor, D H. (2005). Events. In E. Craig (Ed.), *The Shorter Routledge Encyclopedia of Philosophy,* London: Routledge.

Merriam-Webster. (n.d.). Innovation. In *Merriam-Webster Dictionary.* Retrieved April 12, 2018, from https://www.merriam-webster.com/dictionary/innovation

Merriam-Webster. (n.d.). Science. In *Merriam-Webster Dictionary.* Retrieved June 26, 2024, from https://www.merriam-webster.com/dictionary/science

Miller, P. G., Strang, J., & Miller, P. M. (Eds.), (2010). *Addiction Research Methods,* John Wiley & Sons.

Miller, A. (2005). Objectivity. In E. Craig (Ed.), *The Shorter Routledge Encyclopedia of Philosophy,* London: Routledge.

Mitchell, M. L., & Jolley, J. M. (2009). *Research design explained: Instructor's edition,* Belmont: Wadsworth.

Mitchelmore, M., & White, P. (2004). Abstraction in Mathematics and Mathematics Learning. *International Group for the Psychology of Mathematics Education.*

Moore, J. C., Stinson, L. L., & Welniak, E. J. (2000). Income measurement error in surveys: A review. *Journal of Official Statistics-Stockholm,* 16(4), 331-362.

Morreau, M. (2009). The hypothetical syllogism. *Journal of Philosophical Logic,* 38(4), 447-464.

Moser, P. K. (2005). A priori. In E. Craig (Ed.), *The Shorter Routledge Encyclopedia of Philosophy,* London: Routledge.

Mousseau, M. C. (2003). Parapsychology: Science or pseudo-science. *Journal of Scientific Exploration,* 17(2), 271-282.

Mulaik, S. A. (1986). Toward a synthesis of deterministic and probabilistic formulations of causal relations by the functional relation concept. *Philosophy of Science,* 53(3), 313-332.

Mumford, S., & Anjum, R. L. (2010). A powerful theory of causation. In *The Metaphysics of Powers,* 151-167, Routledge.

Neuber, M. (2014). Is logical empiricism compatible with scientific realism?. In *European philosophy of science–philosophy of science in Europe and the Viennese heritage,* 249-262, Springer.

Ney, A. (2012). Neo-positivist metaphysics. *Philosophical Studies,* 160, 53-78.

Nickerson, R. S. (2011). Logic and reasoning. In M. A. Runco & S. R. Pritzker (Eds.), *Encyclopedia of Creativity,* Academic Press.

Nicolaou, A. I. (2011). A dialectic on methodologies in AIS research. *International Journal of Accounting Information Systems*, 2(12), 81-83.

O'Connor, C. L. & A. C. Carol. (1995). A case study for McGregor's motivation theories X and Y. *JONA*, 25 (12).

Oliver, A. (2005). Facts. In E. Craig (Ed.), *The Shorter Routledge Encyclopedia of Philosophy*, London, Routledge.

Onwuegbuzie, A. J., & Daniel, L. G. (2003). Typology of analytical and interpretational errors in quantitative and qualitative educational research. *Current Issues in Education*, 6(2).

Oswald, M. E., & Grosjean, S. (2012). Confirmation bias. In *Cognitive illusions*, 91-108, Psychology Press.

Outhwaite, W., & Turner, S. (Eds.). (2007). *The Sage Handbook of Social Science Methodology*. London: Sage

Oxford Dictionary (n.d.). Analytical. In *Oxford Dictionary*. Retrieved October 19, 2023, from https://www.oed.com/dictionary/analytical_adj?tab=meaning_and_use#4218161

Oxford Dictionary (n.d.). Classification. In *Oxford Dictionary*. Retrieved February 21, 2021, from https://www.oed.com/dictionary/classification_n?tab=meaning_and_use#9307623

Oxford Dictionary (n.d.). Dialectic. In *Oxford Dictionary*. Retrieved June 3, 2022, from https://www.oed.com/dictionary/dialectic_n1?tab=meaning_and_use #6920761

Oxford Dictionary, (n.d.), Globalization. In *Oxford Dictionary*. Retrieved November 28, 2023, from https://www.oed.com/dictionary/globalization_n?tab=meaning_and_use#3012225

Oxford Dictionary, (n.d.), Induction. In *Oxford Dictionary*. Retrieved September 6, 2023, from https://www.oed.com/dictionary/induction_n?tab=meaning _and_use#529234

Oxford Dictionary, (n.d.), Phenomenology. In *Oxford Dictionary*. Retrieved January 7, 2024, from https://www.oed.com/dictionary/phenomenology_n?tab=meaning_and_use#30749550

Oxford Dictionary, (n.d.), Rationality. In *Oxford Dictionary*. Retrieved June 19, 2022, from https://www.oed.com/dictionary/rationality_n?tab=meaning_and_use#26733032

Oxford Dictionary, (n.d.), Relatedness. In *Oxford Dictionary*. Retrieved August 15, 2023, from https://www.oed.com/dictionary/relatedness_n?tab=meaning _and_use#25955938100

Oxford Dictionary, (n.d.), Terminology. In *Oxford Dictionary*. Retrieved August 9, 2020, from https://www.oed.com/dictionary/terminology_n?tab=meaning _and_use#18960688

Özlem, D. (1994). *Mantik*. Istanbul: Anahtar Yayinlari.

Papageorgiou, K., & Lekkas, D. (2020). Verification in theory and in the sciences. *Epistēmēs Metron Logos*, (3), 25-48.

Papineau, D. (2016). Naturalism, In E. N. Zalta, (Ed.), *The Stanford Encyclopedia of Philosophy* (Winter 2016 Edition), Retrieved July 24, 2019, from <https://plato.stanford.edu/archives/win2016/entries/naturalism/>.

Park, K., Daston, L. J., & Galison, P. L. (1984). Bacon, Galileo, and Descartes on imagination and analogy. *Isis*, 75(2), 287-289.

Patton, M. Q. (2014). *Qualitative research & evaluation methods: Integrating theory and practice.* Sage publications.

Peng, K., & Nisbett, R. E. (1999). Culture, dialectics, and reasoning about contradiction. *American Psychologist*, 54(9), 741.

Pereira Jr, A., & Ricke, H. (2009). What is consciousness?: towards a preliminary definition. *Journal of Consciousness Studies*, 16(5), 28-45.

Peters, U. (2020). What is the function of confirmation bias?, *Erkenntnis*, 1-26.

Pigliucci, M., & Boudry, M. (2013). *Philosophy of Pseudoscience: Reconsidering the Demarcation Problem.* University of Chicago Press.

Platt, J. (2007). Case Study. In W. Outhwaite & S. Turner (Eds.). *The Sage Handbook of Social Science Methodology.* London: Sage.

Popper, K. R. (1972). *Objective Knowledge* (Vol. 360). Oxford: Oxford University Press.

Popper, K. R. (1974). In *Does science teaching need history and philosophy of science?.* Slezak, P. (1994). Science teaching: The role of history and philosophy of science (21–38). New York: Routledge.

Preacher, K. J., & Hayes, A. F. (2008). Asymptotic and resampling strategies for assessing and comparing indirect effects in multiple mediator models. *Behavior Research Methods*, 40(3), 879-891.

Pritchard, D. (2014). *What is this thing called knowledge?,* New York: Routledge.

Proctor, R. N., & Schiebinger, L. (2008). Agnotology: *The making and unmaking of ignorance.* Stanford University Press.

Quine, W.V.O. (1951). Two dogmas of empiricism. *The Philosophical Review*, 60(1), 20-43.

Radford, (n.d.). Analytical Reasoning, In *Radford University Website.* Retrieved October 19, 2023, from https://www1.radford.edu/content/cobe/innovation-analytics/analytics/career-prep/report-e.html#par_text_1039098517

Ragin, C. C. (2007). Comparative methods. In W. Outhwaite & S. Turner, (Eds.), *The Sage Handbook of Social Science Methodology.* London: Sage.

Raleigh, T. (2022). Philosophy of perception and liberal naturalism. In *The Routledge Handbook of Liberal Naturalism*, 299-319, Routledge.

Ranis, G., Stewart, F., & Ramirez, A. (2000). Economic growth and human development. *World Development*, 28(2), 197-219.

Ranis, G. (2004). *Human development and economic growth.* Discussion Paper No. 887, Yale University.

Reed, B. (2012). Fallibilism. *Philosophy Compass*, 7(9), 585-596.

Rescher, N. (2003). *Epistemology: An introduction to the theory of knowledge.* Albany: State University of New York.

Rey, G. (2005). Concepts. In E. Craig (Ed.), *The Shorter Routledge Encyclopedia of Philosophy,* London: Routledge.

Ritzer, G., & Ryan, J. M. (Eds.). (2011). *The Concise Encyclopedia of Sociology.* John Wiley & Sons.

Rivera, R. (2020). *Principles of Managerial Statistics and Data Science.* New Jersey: John Wiley & Sons.

Robbins, S. P & M. Coulter. (2012). *Management,* (11th. Ed.), New Jersey: Pearson Education.

Rodrigues, F. H., & Abel, M. (2019). What to consider about events: A survey on the ontology of occurrents. *Applied Ontology,* 14(4), 343-378.

Rogelberg, S. G., & Olien, J. L. (2014). Research design. In C. L. Cooper, N. Lee, & A. Farrel (Eds.), *Wiley Encyclopedia of Management,* 1-3. John Wiley & Sons.

Ropohl, G. (1997). Knowledge types in technology. In M. J. De Viries & A. Tamir (Eds.), *Shaping concepts of technology from philosophical perspective to mental images.* 65-72. Springer, Dordrecht.

Root, M. (2007). Community-Based Research. In W. Outhwaite & S. Turner (Eds.), *The Sage Handbook of Social Science Methodology.* London: Sage.

Rosenberg, A. (2008). *Philosophy of social science.* Westview Press.

Rosenberg, A. (2012). *The Atheist's Guide to Reality: Enjoying Life without Illusions.* W. W. Norton & Company.

Rosenkrantz, G. (2012). Ontological categories. *Contemporary Aristotelian Metaphysics,* 83-93.

Rosenstein, L. D. (2019). *Research design and analysis: A primer for the non-statistician.* John Wiley & Sons.

Rothchild, I. (2006). Induction, deduction, and the scientific method. In *Society for the Study of Reproduction,* 1-11.

Rouse, J. (2007). Naturalism and scientific practices: A concluding scientific postscript. *Naturalized Epistemology and Philosophy of Science,* 61-86.

Roy, T. *Logical Validity and Soundness.* tonyroyphilosophy.net

Russo, F. (2017). Causation and Correlation in Medical Science: Theoretical Problems. In Schramme, T. and Edwards, S. (Eds.), *Handbook of the Philosophy of Medicine,* 839-849, Springer.

Salmon, M. H., & Glymour, C. (1999). *Introduction to the Philosophy of Science.* Hackett Publishing.

Sankey, H. (2012). Scepticism, relativism and the argument from the criterion. *Studies in History and Philosophy of Science Part A,* 43(1), 182-190.

Sawyer, R. K. (2007). Simulating Complexity. In W. Outhwaite & S. Turner (Eds.), *The Sage Handbook of Social Science Methodology.* London: Sage.

Schmidt, F. (2010). Detecting and correcting the lies that data tell. *Perspectives on Psychological Science,* 5(3), 233-242.

Schnake, M. E., & Dumler, M. P. (2003). Levels of measurement and analysis issues in organizational citizenship behaviour research. *Journal of Occupational and Organizational Psychology,* 76(3), 283-301.

Schneider, J. K., & Deenan, A. (2004). Reducing quantitative data errors: tips for clinical researchers. *Applied Nursing Research,* 17(2), 125-129.

Schwandt, T. A. (2001). *Dictionary of qualitative inquiry,* Thousand Oaks, California: Sage.

Schwandt, T. A. (2015). Epistemology. In G. Ritzer (Ed.), *The Blackwell Encyclopedia of Sociology,* John Wiley & Sons.

Science Council. (n.d.). *Our definition of science.* Retrieved June 26, 2024, from https://sciencecouncil.org/about-science/our-definition-of-science/

Segel, E., & Heer, J. (2010). Narrative visualization: Telling stories with data. *IEEE Transactions on Visualization and Computer Graphics,* 16(6), 1139-1148.

Sekaran, Uma (2003), *Research Methods for Business: A Skill-Building Approach,* New York: John Wiley & Sons.

Settle, T. (1971). The rationality of science versus the rationality of magic. *Philosophy of the Social Sciences,* 1(2), 173-194.

Shang, H. (2006). Actuarial science: Theory and methodology. In H. Shang (Ed.), *Actuarial Science: Theory and Methodology,* Higher Education Press and World Scientific.

Shrum, L. J. (2008). Selective Perception and Selective Retention. *The International Encyclopedia of Communication.*

Shustov, B. M., & Rykhlova, L. V. (2009). Asteroid-comet impact hazard: New approaches. *Herald of the Russian Academy of Sciences,* 79(4), 317-323.

Simon, H. A. (1955). "A behavioral model of rational choice." *Quarterly Journal of Economics,* 69(1), 99-118.

Smith, D. W. (2018). Phenomenology, In E. N. Zalta (Ed.), *The Stanford Encyclopedia of Philosophy,* Retrieved 19 June, 2021, from https://plato.stanford.edu/archives/sum2018/entries/phenomenology/

Smith, E. E., Langston, C., & Nisbett, R. E. (1992). The case for rules in reasoning. *Cognitive Science,* 16(1), 1-40.

Smith, J. K. (2005). Inflation targeting and core inflation. *Canadian Journal of Economics/Revue canadienne d'économique,* 38(3), 1018-1036.

Smith, K. B. (2002). Typologies, taxonomies, and the benefits of policy classification. *Policy Studies Journal,* 30(3), 379-395.

Smithson, M. (1985). Toward a social theory of ignorance. *Journal for the Theory of Social Behaviour,* 15(2), 151-172.

Snidal, D. (1985). The game theory of international politics. *World Politics: A Quarterly Journal of International Relations,* 25-57.

Snijkers, G., Haraldsen, G., Jones, J., & Willimack, D. (2013). D*esigning and conducting business surveys,* John Wiley & Sons.

Sprenger, J. (2016). Confirmation and induction. *Oxford Handbook of the Philosophy of Science,* 185-209.

Stanley, J., & Williamson, T. (2001). Knowing how. *The Journal of Philosophy,* 98(8), 411-444.

Stephens, J., Barton, J., & Haslett, T. (2009). Action research: its history and relationship to scientific methodology. *Systemic Practice and Action Research,* 22(6), 463.

Strawson, G. (2005). Free will. In E. Craig (Ed.), *The Shorter Routledge Encyclopedia of Philosophy,* London: Routledge.

Tanaka, S. (2019). *History without chronology,* Lever Press.

Teller, E., & Brown, A. (1962). *The legacy of Hiroshima.* Doubleday. New York.

Tseng, Y. C., Tsai, C. Y., Hsieh, P. Y., Hung, J. F., & Huang, T. C. (2014). The relationship between exposure to pseudoscientific television programmes and pseudoscientific beliefs among Taiwanese university students. *International Journal of Science Education, Part B,* 4(2), 107-122.

Thagard, P. R. (1978, January). Why astrology is a pseudoscience. In PSA: Proceedings of the Biennial Meeting of the Philosophy of Science Association. *Philosophy of Science Association,* (1), 223-234.

Tinati, R., Halford, S., Carr, L., & Pope, C. (2014). Big data: methodological challenges and approaches for sociological analysis. *Sociology,* 48(4), 663-681.

Tomal, D. R. (2010). *Action research for educators.* Rowman & Littlefield Publishers.

Tomaszewski, C. M. (2016). The Principle of Sufficient Reason Defended: There Is No Conjunction of All Contingently True Propositions. *Philosophia,* 44(1), 267-274.

Tufte, E. R. (2001). *The visual display of quantitative information.* Graphics Press.

Turk, M. (2003). *Kuresellesme surecinde isletmelerde bilgi yonetimi.* Istanbul: Turkmen Kitabevi.

Ultee, W. C. (2015). Problem selection in the social sciences: methodology. In Wright, J.D. (Ed.), *International encyclopedia of the social & behavioral sciences* (2nd ed.), 49-55. Amsterdam: Elsevier.

Vrij, A. (2008). *Detecting lies and deceit: Pitfalls and opportunities.* John Wiley & Sons.

Wallis, R. (1985). Science and pseudo-science. *Social Science Information,* 24(3), 585-601.

Wardle, S. G., Taubert, J., Teichmann, L., & Baker, C. I. (2020). Rapid and dynamic processing of face pareidolia in the human brain, *Nature Communications,* 11(1), 1-14.

Watkins, L., M., (n.d.). Employee Motivation and Job Satisfaction. In *Family Owned and Corporate Businesses,* Retrieved September 22, 2023, from https://www.mckendree.edu/academics/scholars/issue9/watkins.htm

Wayne, M. Genoves, M., & Shimony, A. (2021) Bell's Theorem, In E. N. Zalta, (Ed.), *The Stanford Encyclopedia of Philosophy,* Retrieved May 20, 2022, from https://plato.stanford.edu/entries/bell-theorem/

Weathington, B. L., Cunningham, C. J., & Pittenger, D. J. (2012). *Understanding Business Research.* John Wiley & Sons.

Webster, M., & Sell, J. (2007). Theory and experimentation in social sciences. In W. Outhwaite & S. Turner (Eds.), *The Sage Handbook of Social Science Methodology.* London: Sage.

Westphal, K. R. (2017). Empiricism, pragmatic realism, and the a priori in Mind and the world order. *Pragmatism in transition: Contemporary perspectives on CI Lewis,* 169-198.

White, J. & Lintzeris, N. (2010). Psychopharmacology. In P. G. Miller, J. Strang, & P. M. Miller (Eds.), *Addiction Research Methods.* John Wiley & Sons

Winograd, T., & Flores, F. (1986). *Understanding computers and cognition: A new foundation for design* (Vol. 335). New Jersey: Addison-Wesley.

Winter, R., & Munn-Giddings, C. (2001). *A handbook for action research in health and social care.* Routledge.

Winther, R. G. (2012). Interweaving categories: Styles, paradigms, and models. *Studies in History and Philosophy of Science Part A,* 43(4), 628-639.

Worrall, J., (2005). Philosophy of science. In E. Craig (Ed.), *The Shorter Routledge Encyclopedia of Philosophy,* London, Routledge.

Vander Nat, A. (2010). *Simple formal logic: with common-sense symbolic techniques.* Routledge.

Von Bertalanffy, L. (1950). An outline of general system theory. *The British Journal for the Philosophy of science,* 1(2), 134-165.

Yang, Y. (2001). *Basic concepts of research in economics* (Chapter 1), https://www.csus.edu/indiv/y/yangy/145ch1.htm

Yildirim, A., Simsek, H., (2011) *Sosyal bilimlerde nitel arastırma yontemleri,* Ankara: Seckin Yayincilik.

Zwart, H. (2021). Dialectical Materialism. In *Continental Philosophy of Technoscience.* 67-109. Springer International Publishing.

Image sources

Source of Figure 16. Simulation of Typhoon Mawar (2005): https://commons.wikimedia.org/wiki/File:Typhoon_Mawar_2005_computer_simulation.gif

Source of Figure 17. A vector illustration of tectonics: https://pixabay.com/vectors/dorsal-tectonics-plates-geology-6801801/

Smiley figures in quotation boxes: https://pixabay.com/vectors/smiley-happy-face-smile-lucky-559124/

Data sources

https://www.ssa.gov/policy/docs/chartbooks/fast_facts/2016/fast_facts16.html#page5

https://worldpopulationreview.com/world-cities

United Nations World Population Prospects, 2019

INDEX

www.ingramcontent.com/pod-product-compliance
Lightning Source LLC
LaVergne TN
LVHW020414220926
848484LV00018B/2579/J